Y0-BZK-085

Robert Koch

A Life
in Medicine
and Bacteriology

Robert Koch

A Life
in Medicine
and Bacteriology

Thomas D. Brock

With a new foreword

ASM PRESS

Washington, D.C.

Copyright © 1999 American Society for Microbiology
1325 Massachusetts Avenue NW
Washington, DC 20005-4171

Previous edition copyright © 1988 Science Tech Publishers; copyright assigned to Thomas D. Brock in 1994. All rights assigned to ASM by Thomas D. Brock in 1998.

Library of Congress Cataloging-in-Publication Data

Brock, Thomas D.
   Robert Koch : a life in medicine and bacteriology / Thomas D. Brock.
      p.   cm.
   "With a new foreword."
   Originally published: Madison, Wis.: Science Tech Publishers; Berlin, New York: Springer-Verlag, 1988.
   Includes bibliographical references and index.
   ISBN 1-55581-143-4 (hbk.)
   1. Koch, Robert, 1843–1910.   2. Microbiologists—Germany—Biography.   I. Title.
QR31.K6B76   1998
616'.014'092—dc21
[B]   98-43123
                                                                                                    CIP

All Rights Reserved
Printed in the United States of America

10 9 8 7 6 5 4 3 2 1

Contents

Foreword

Ten years have passed since Dr. Thomas Brock gave us this first major English-language biography of Robert Koch in 1988. Written by a distinguished microbiologist rather than a professional historian or biographer, the book has nonetheless stood the test of time. Indeed, though Brock did not work with unpublished archival sources or dissertations, his command of the published sources is extremely impressive, especially his use of the German-language biographies of Koch that are otherwise inaccessible to most English-speaking readers. Thus, although Brock did not make use of archival materials (many of which, in East Berlin, may have become available to scholars only after the fall of the Berlin Wall), his analysis in many ways opens up or foreshadows important themes that historians have since developed in more depth. His bibliography and notes are very well arranged so that this biography is useful to the average reader *and* an essential starting point for any historians studying Koch from this point on.

In his lifetime, Robert Koch did more to singlehandedly advance the world's understanding of microbes as causes of disease than any other man, with the exception only of his great French rival Louis Pasteur. Koch and his students created almost from scratch the majority of the techniques necessary for any modern study of bacteria, including microphotography of the organisms, staining procedures, and solid culture media, which allowed reproducible pure cultures and quantification of bacterial numbers for the first time. They also identified the microorganisms that cause anthrax, wound infections, tuberculosis, diphtheria, cholera, and many other major infectious diseases. Koch had the talents of a first-rate researcher: he was a keen observer, an ingenious technical innovator, and extremely persistent and single-minded in pursuit of his goal. Yet, being trained from the outset of his career as a practicing medical doctor, Koch never lost sight of the practical benefit to human health that was implicit in his work, for instance, devoting much time to the study of effective water filtration systems after the 1892 Hamburg cholera epidemic. This is not a mere academic point; Brock reminds us (p. 3) that "water filtration has probably saved more lives than immunization and chemotherapy combined." Brock is thorough, and his deep admiration for Koch is obvious, as he documents in detail all of Koch's important contributions, without which it is difficult to imagine the existence of bacteriology as we know it.

Yet this book stands out from a great many hagiographic biographies of the past. With figures of the stature of Koch or Pasteur, one rarely finds a study that can document the awesome contributions of the scientist without setting him up as a hero figure, a giant with no human qualities. Brock, more in tune with the needs of our own time, sees that Koch's human qualities make an equally fascinating part of the story. An important thread in this narrative of Koch's career is to show how an "eager amateur" country doctor, experimenting on microbes in his spare time, went on to become "an imperious and authoritarian father figure whose influence on bacteriology and medicine was so strong as to be downright dangerous." And the point, Brock emphasizes, is not just to understand Koch's personality, but to better understand science itself, including "the origins of the cult of personality in research" (p. 4). It might be added that, for a new generation of scientists in training, it is much more valuable as a model for study to have a real, complex human being, including his mistakes and excesses. By comparison, a too-perfect hero figure will always leave most of us doubting that we can measure up, especially in the face of the messy enterprise that laboratory science can be on a day-to-day basis.

I shall not attempt an exhaustive summary of the past decade's historical work on Koch, but I shall briefly indicate some of the most important works and what contribution each has made. For the reader who wants more detail, the bibliographies of those works will provide a more than adequate guide to the literature. Subjects that have received excellent and fascinating treatment include the rivalry between the Koch and Pasteur schools (and the two men personally), the pleomorphism-monomorphism debate, the greater complexity of Koch's postulates than at first meets the eye, the larger political and cultural resonances of the bacteriological revolution (especially in Germany), the tuberculin discovery and its relation to the creation of Koch's Institute for Infectious Diseases, and finally the difficulties of the historian's task in adequately assessing a scientific giant without merely re-creating the mythic hero figure that the scientist often himself began constructing during his own lifetime.

Koch disagreed with Pasteur on a number of important issues in bacteriology. Brock tells us that in the Pasteur-Koch controversy, Koch could at times be so personally vicious as to be shocking. We learn, however, that the controversy "was certainly rooted, in part, in the French-German antagonism that still festered as an aftermath of the Franco-German war." This has also been documented for Pasteur in the most recent work. Brock points out (p. 176–177) that differences in *style* between the Koch and Pasteur schools, also "rooted in national characteristics," but with "more deep-seated significance" scientifically, contributed to the animosity and misunderstanding as well. These themes have been explored and developed in great depth by Andrew Mendelsohn (originally in his Princeton Ph.D. dissertation, which is soon to appear as a book), who characterizes the French and German schools as two fundamentally different "cultures of bacteriology." He argues that Pasteur's more ecological approach to microorganisms was rooted in an agricultural French context in which the "economy of nature" (and hence the role of microorganisms in that big picture) was a primary cultural motif. Indeed, Pasteur's entire work with microbes began through fermentation, their positive functions such as winemak-ing suggesting the ecological necessity of microbes. By contrast, Koch's view of bacteria was medically rooted from the beginning, his surgical experience in the Franco-Prussian war; microbes were pathogens, ex-clusively negative agents to be eliminated if at all possible. This fun-damental difference in "culture" produced different methods of work-ing with bacteria, e.g., Koch's early insistence on working with pure cultures as opposed to the Pastorians' preference for working with liq-uid cultures that, by nature, were mixed populations. But Mendelsohn shows that the deep philosophical difference continued to be mani-fested in the French and German "schools" and as the new science of bacteriology was transformed over several decades.

Another important contribution to this aspect of Koch's work has come from University of Toronto historian Pauline Mazumdar. She shows convincingly that an important early source of Koch's basic op-position to the Pastorians was his deep epistemological commitment to monomorphism, the idea that microbes, like all other organisms, come only in discrete species. Carl von Nägeli in Munich and his students championed the opposite theory, pleomorphism, that microbes can undergo such a wide range of mutability under different environmental conditions that the morphologically different types are almost all interconvertible. Mazumdar shows that these views on microorganisms were manifestations of very long-standing opposed views of nature that she calls the "Linnaeans" and the "Unitarians." Koch emulated the work of the botanist Ferdinand Cohn, with whom he shared the view that bacteria came in true Linnaean species, and thus he was opposed to Nägeli's ideas as soon as he heard them. It was Koch's deep a priori belief in this view of nature that led him to think that photography of the organisms would be an important contribution to their study. Mazumdar suggests that Koch's insistence, as early as 1878 (long before there was any proof of such an idea), that one bacterial species must be the cause of one and only one disease also grew out of this basic difference in beliefs about nature. In his 1878 paper on wound infections, he used this as a new dimension to his definition of bacterial species: the disease that an organism caused was one of the chief features that could be used to define which species of microbe it must be. Brock pointed out many years ago in his book *Milestones in Microbiology* (p. 100–101) that Koch was exhibiting theory-laden observation when he emphasized tiny differences between, e.g., micrococci and indeed that the researcher was begging the question at issue in imagining "that the minor differences he saw were significant. . . . He wanted these organisms to be different" so that his theory would be verified. Brock concludes, "Fortunately Koch was right on this point, but there was no a priori reason why he should have been right, and so we must conclude that he was lucky." Thus, the widespread opposition Koch faced from other doctors and students of microbes was not at first due to their being unaware of his evidence. Rather, it was precisely because they saw that he did not have the evidence for his strong one bacterium-one disease claim that his theory was opposed and the theories of Nägeli and Max von Pettenkofer retained wide support into the 1890s.

So why should this contribute to antagonism between Koch and the Pasteur school as well? Mazumdar points out that the pleomorphist camp used exclusively liquid cultures, as did everybody else prior to Koch's invention of solid media. But Koch was convinced that it was that fact above all else that caused confusion between (what he was sure were) different separate species in a mixed culture and (what his opponents interpreted as) the different stages in the interconverting life cycle of pleomorphic microbes. When Pasteur first announced the discovery of attenuated virulence in a microbe in 1880, this reminded Koch all too much of the pleomorphists' claims that such a fundamental defining property of the organism as its ability to cause a disease was a mutable thing. And since the Pastorians, like Nägeli and his followers, still used liquid media for their cultures, this clinched the case in Koch's mind: the Pasteur school was highly suspect of being as fundamentally misguided about stable, unchanging bacterial species as were the pleomorphists.

Of course, we now know that Koch was wrong in thinking that stable species were incompatible with quite significant genetic mutability within a species of microorganisms. Indeed, some would argue that the discovery of such important phenomena as the variation among "smooth" and "rough" forms of pneumococci (and the resultant path to the double helix) may have been held up for decades by the extent to which Koch's monomorphist dogma held sway, once he triumphed with the discovery of the causes of spectacular diseases such as cholera and tuberculosis. Bacteriologist Ludwik Fleck made such an argument in the 1930s. Brock also cites Philip Hadley and shows that disagreement over this point had surfaced among American bacteriologists even in the last years of Koch's life. Yet the slow waning of Koch's dogmatically extreme monomorphist view was still to create trouble, even for American researchers, as Harris Coulter has shown in the case of Arthur Isaac Kendall. Clearly, this is a prime example of what Brock intimated when he warned of the excessive influence of the "cult of personality" that grew up around Koch. While his insights about the use of pure cultures were brilliant, his a priori bias led him to mistakenly think this must mean that Pasteur's discovery of variability among microbes was illusory. For the next several decades, this necessitated the creation of epistemological wastebasket categories such as "involution forms," into which observations could be banished when they seemed at odds with the monomorphist paradigm. If an overriding commitment to his belief had the benefit of driving Koch through the years of hard work necessary to isolate the causes of major human woes, it simultaneously had the effect, proportional to his success, of calcifying the research and preserving Koch's mistakes, at least for several decades. It also, as much as nationalistic feelings, may have contributed to driving a deep wedge of suspicion between him and the only other group of workers with whom he could have collaborated as equals. Such is a common feature of research that should give us pause.

One of the single most important new historical contributions on Koch is Richard Evans's masterly analysis of the 1892 Hamburg cholera epidemic, in which Koch and his new bacteriology clashed with the sanitarian theory of Max von Pettenkofer. Pettenkofer's theory was based on exhaustive study of local conditions of soil and climate and emphasized that the cholera germ was only one ingredient needed to produce the disease. Only when the bacterium came in contact with the soil under specific conditions related to the underground water table could the actual cholera *poison* be generated. Thus, in Pettenkofer's theory, which had enjoyed wide respect in medical circles for 20 years by 1892, the bacteria getting into drinking water could in no way transmit the disease. In the epidemic of 1892, it was noticed that the rate of cholera in the immediately adjacent, downstream city of Altona was negligible compared to the rate in Hamburg. Altona had a sand filtration system for its water supply, so to Koch and his supporters it seemed obvious that drinking water must be the main means of transmission of the disease. Historically, one of the things calling out for explanation has been how anyone in Hamburg could have resisted such compelling epidemiological evidence and not immediately thrown out the Pettenkofer theory in favor of the Koch theory that bacterium equals disease. Yet the epidemic reached very serious proportions indeed, claiming nearly 10,000 lives in 6 weeks before Koch's views came to dominate. The power of Evans's close-up history is in explaining which forces in the medical community and government of Hamburg lent support to Pettenkofer's theory and why. Furthermore, Evans shows that larger political tensions between Hamburg and the German Empire (with Prussia the dominant state), over "federal" intervention in local matters, exacerbated the reasons why local Hamburg officials would be opposed to Koch, an official of the Prussian bureaucracy.

The famous historian of medicine Erwin Ackerknecht first suggested 50 years ago that anticontagionist, sanitarian theories of epidemic disease would tend to be supported by 19th century classical liberals (including free-trade advocates) while contagionist theories would be more likely to be supported by political conservatives. Why? Because contagionist theory implied the need for more centralized government authority and interference in local affairs to enforce quarantines and disinfection measures. Free traders, especially merchants and businessmen whose livelihood depended upon the free and cheap movement of goods across state and national boundaries, stood to lose most if quarantines were imposed because of epidemics. In the 19th century, this group tended toward reformist, liberal politics. More extreme liberals, such as the famous pathologist Rudolf Virchow, insisted that social reforms for the underfed, overworked poor who lived in unclean conditions were the only real cure for epidemic diseases. Virchow and his supporters would always be highly suspicious of germs as any kind of true causative agents, recognizing that the easiest way for a conservative government (such as that of Prussia or, after 1871, the Prussian-dominated Empire) to avoid expensive and democratizing social reforms was to blame epidemics entirely upon a germ from without, thus avoiding issues of poverty and inequality altogether and insisting that all that was needed was quarantine and disinfection. These far-left liberals were not surprised that the imperial government in Berlin supported Koch. And they were skeptical of Pettenkofer's theory for allowing any role for a germ at all. Yet, compared to Koch's Prussian state antigerm bureaucracy, which gave the germ total causative blame, Pettenkofer's theory still appealed to liberals because it did at least emphasize the importance of local conditions in creating the actual poison that caused disease. Thus, it implied that local medical officials, not far-off Prussian bureaucrats trying to pass sweeping uniform policies and force them on all German cities under all conditions, were by far the most appropriate people to decide how best to deal with epidemics. Practically speaking, sanitarian theories did actually greatly reduce overall mortality from epidemics because of their emphasis on building sewers and public water supplies (though not necessarily with filtration) and on improving nutrition and general living conditions. Thus, in England and in many areas of Germany, sanitary theory was credited with actually solving the problems, without the need for germs as central players.

If all this were not enough to make Pettenkofer's theory more popular with most local medical officials, Evans shows that Hamburg was an even more special case and proves that this political context was a very important reason why Hamburg, alone among German cities by 1892, experienced a severe cholera outbreak that year. Hamburg's government had been run for centuries by the mercantile class, as the merchants had basically made the wealth of this trading port since medieval times, when it first became a free and independent city-state within the Hanseatic League. In its liberal, free-trade policies and culture, the city had long been known as the "most English city on the

Continent," let alone in Bismarck's conservative German Empire. The mercantile ruling class selected the medical officials, and the doctors most likely to become public servants were those who saw their interests most closely tied to those of the wealthy merchants. Thus, the entire history and culture of the city militated against quarantine policies and the havoc they caused in disrupting trade, and Hamburg held out longer than any other German city against centralized control of local medical policies by Berlin. In 1892, huge numbers of eastern Europeans were passing through Hamburg to board ships for emigration abroad, especially to America. At that time, if ever, city officials would be loath to impose a lengthy quarantine, stopping the flow of this highly profitable cargo. Yet it was just the flow of these poor emigrants that was bringing the cholera bacillus from the east and depositing it in the sewers, the river, and the harbor of Hamburg. Because one of the most pointed differences between the Koch and Pettenkofer approaches was in whether epidemic disease poisons could be spread by drinking water supplies, the kind of epidemic Hamburg (or Pettenkofer's Munich) was most unprepared for, despite other intelligent sanitarian measures, was an epidemic of a water-borne disease such as cholera or typhoid. Since centralized Berlin policy on germs, as dictated from Koch's lab, was enforced through almost the entire remainder of the Empire, only Hamburg fell victim to cholera that year. Needless to say, Koch was sent to investigate, and his eventual triumph over the 1892 epidemic was a crippling blow to Hamburg's continued economic and political independence from Prussian domination. It is not possible to predict simply and unequivocally that any given doctor would support or oppose contagionist theory and policy based solely on his basic political views. Nevertheless, the integral nature of political history in understanding the fortunes of the germ theory of disease never came through more clearly than in Evans's story of *Death in Hamburg*.

Brock's biography makes eminently clear that Koch understood how crucial the support of the imperial and Prussian state politicians in Berlin was for the advancement of his career and the spread of his ideas. And the relationship was a true symbiosis. For, in the wake of worldwide fame for French science that came after Pasteur's triumphs with anthrax vaccine in 1881, the prestige of German science was at stake. If the memory of the Franco-Prussian war 10 years before were not still clearly in everyone's mind, Pasteur was deeply embittered and publicly campaigned all through these years for more support for

French science, insisting that France had fallen behind the state support the Germans gave to science and that this was an important reason for her defeat by the "Prussian chancre." Thus, supporting Koch and trumpeting his triumphs as the German answer to Pasteur were priorities for Berlin. The famed "race" between the German and French teams to find the germ of cholera in 1883 and the declaration of Koch's *Vibrio cholerae* in 1884 as a triumph for German science must be seen in this context. There are two reasons. First, several other investigators had observed the cholera bacillus before Koch (the Italian Pacini is officially credited with having first seen it in 1854), so presumably more conclusive proof of a causative link should account for the contemporary trumpeting of credit for Koch. But the second point is exactly this: Koch was *unable* to fulfill the criteria for proof of causation that were to be announced in that same year and later became enshrined in our microbiology textbooks as Koch's postulates. The most crucial missing link was the inability to infect an animal model with the bacterium and cause the disease. Thus, the Pettenkofer school, and many outside Germany as well, was highly skeptical of whether Koch's bacillus was any more proven to cause cholera than any of the other numerous intestinal bacteria for which the claim had been made before. Anticontagionists recalled in particular an episode in 1849 when British researchers claimed to have shown that a fungus was the cause of cholera, only to have it shown within the year that the organism in question was a common mold contaminant. Why, in this context, Koch's *Vibrio* came to be so widely celebrated makes much more sense if we recall that the Berlin government was the most vocal advocate of that view, treating Koch and his team as national heroes upon their return from India and arranging for Koch to be publicly greeted by the German emperor.

The French were not the only intended targets of this orchestrated propaganda for the superiority of German science. As Evans points out, 1884 was the year that saw the beginning of the imperial powers' "scramble for Africa." And the furious competition to conquer disease in the name of science ran neck-and-neck over the next 3 decades with the race among the Germans, French, British, and others to colonize territory in the name of civilization. The link was twofold: a propaganda war to justify imperialist expansion on one hand, combined with the need for science to control aggressive tropical diseases so that large numbers of Europeans, especially troops, could live in Africa on the other. By 1896, Koch had begun to shift his major research interest to the tropical diseases of Africa. That Koch's scientific interest moved in this direction is not to be doubted. But again we must recall that it was only the support of the German government that made *possible* Koch's intensive full-time work in bacteriology after 1880. So perhaps it is not so surprising, and even may be instructive for modern high-budget science, to study the degree to which he who paid the piper chose in this case to call the tune. The racist and political roots of this, along with the science, are explored in depth in recent work by Heidelberg historian Wolfgang Eckart. Eckart shows, for instance, that trials of chemotherapeutic agents against the diseases were much easier to accomplish in the African colonial setting, with nonwhites as experimental subjects, than would ever have been possible in European labs.

To return to the issue of Koch's postulates, this is an area in which very interesting work has been done recently as well. While Koch was willing to deploy these rules in a more rigid form for publicity purposes, we learn from Brock's discussion (p. 180–182) that, from the very beginning, Koch understood that the situation was more problematic. He was convinced that his vibrio was the cause of cholera, for example, and thus that it might still not be possible in every case to fulfill the requirement for reinfecting an animal with the pure culture and reproducing the disease. Thus, from before they were even announced, in the mind of Koch the researcher these rules were not the kind of dogmatic requirements that they went on to become in microbiology textbooks. Historian Victoria Harden of the National Institutes of Health has studied the ongoing conflict in research, especially after the discovery of the viruses (which can almost never be cultivated on nonliving media), between the postulates as a helpful guideline for seeking new disease agents and simultaneously as *obstacles* to new fundamental breakthroughs. Virus researchers have insisted from the earliest days of their work that new versions of the rules must be continually reinvented to take into account the new properties of pathogens that differ from those of the bacteria worked with in the 1880s. Basic disagreements can be caused when two researchers insist on different versions of these postulates as bottom-line criteria, and in no case has this come out more clearly than in the objections of virologist Peter Duesberg that the epidemiological data for human immunodeficiency virus (HIV) are insufficient to prove that it causes AIDS. Harden looks at cases up to and including this one and tries to evaluate the validity of Duesberg's arguments and those of his opponents, such as William Blattner and

Robert Gallo, in light of past historical disputes over which form of Koch's postulates is most reliable. In light of such a history, it is fascinating to reflect on the process by which such a scientific idea, though more flexible in the mind of its creator, can become an obstacle to new discoveries, especially if propagated in too rigid a form in science textbooks. Of course, in allowing the German state to use simplified notions of his work as propaganda tools, Koch himself must have realized his own participation in this process from the beginning.

We see this kind of double-edged nature of patronage again when Brock shows how (p. 198–199) Koch's German government superiors forced him to announce his discovery of tuberculin and its possible curative role for tuberculosis before he thought it scientifically appropriate. The premature announcement was forced on Koch because of the publicity opportunity of making the announcement at the Tenth International Medical Congress. It almost certainly also resulted from the enormous international prestige that came to the Pasteur group in the first years after the development of the rabies vaccine, which led to donations of an enormous sum of money that was used to create the Institut Pasteur in 1888. In the wake of this, Koch's government patrons were planning to create an institute for him in Berlin that would have comparable prestige for cutting-edge research. As soon as his work on tuberculin made it mistakenly seem that it would be an effective therapy for tuberculosis, the negotiations for Koch's new institute became bound up immediately with the potential fame and profit associated with that remedy.

In this area too, recent historical work has also brought new and interesting details of Koch's negotiations with the state bureaucracy to light. Heidelberg scholar Christoph Gradmann has found, in East Berlin archives, detailed government documents showing that the negotiations bogged down in late 1890, but not because Koch was digging in his heels about being forced to announce the discovery prematurely. Koch was trying to strike a deal that would guarantee him a large share of the profits that would accrue from tuberculin sales for the first 6 years. This is in striking contrast with the image of Koch the selfless researcher, whom biographers have been convinced had no real interest in fame or profit. Over the ensuing months, when large-scale trials brought out the fact that tuberculin really had very little therapeutic effect, Koch was forced to back down from his tough stance and accept the creation of the Institute for Infectious Diseases on terms mostly dictated by the German government, since he feared losing all in the public relations debacle over tuberculin. It never became public that, as many tuberculin critics had charged during the months of controversy, Koch hoped to personally profit from the discovery, in addition to getting his institute for the good of humanity.

Here we are faced with something harder to accept into our previous heroic vision of Koch. Despite Brock's thoughtful comments on the larger context of science, politics, and bureaucracy in which Koch worked, here is one area where Brock the microbiologist generously views his subject with the basic faith that Koch was "strongly motivated to excel without regard to fame and fortune," at least in his early years. And yet, working out the role of personal profit in this incredible new field, so important for humanity, is surely an important part of the history of work on human disease. The controversy that followed Selman Waksman and Albert Schatz's discovery of streptomycin, which eventually led to a lawsuit and a court settlement over allocation of profits from that drug, shows that this tension did not go away after the early days of giants like Koch and Pasteur. Surely the recent dispute between Robert Gallo and Luc Montagnier over patent rights resulting from the discovery of HIV shows that it is still a matter very relevant in research. We must realistically include these matters in our picture of Koch to see the full human being and to understand the full relevance of his story for our own times, as well as for the future of scientific research. Paraphrasing from Gerald Geison's recent scholarly and provocative biography of Pasteur, we need a Koch for our times, not only the Koch who has inspired generations of young people to become scientists, but also the more complex person that we know he must have been. This can be done while simultaneously keeping in view the important contributions to science that Koch made so brilliantly. It is a tribute to this book that it has gone so far in that direction without sacrificing the details that make the science itself so compelling, indeed world-changing.

James Strick
Arizona State University

Further Reading
Coulter, Harris. 1994. *Divided Legacy: Medicine and Science in the Bacteriological Era*. North Atlantic, Berkeley, Calif.

Eckart, Wolfgang. 1996. The tropical colony as laboratory: Robert Koch and the fight against sleeping sickness in German East Africa and Togo. Lecture given at conference, Pasteur, Germs and the Bacteriological Laboratory, November 1996, Dibner Institute, Cambridge, Mass.

Evans, Richard. 1987. *Death in Hamburg: Society and Politics in the Cholera Years, 1830–1910*. Oxford University Press, London, United Kingdom.

Fleck, Ludwik. 1979. *The Genesis and Development of a Scientific Fact*. University of Chicago Press, Chicago, Ill. [Reprint of 1935 edition.]

Geison, Gerald. 1995. *The Private Science of Louis Pasteur*. Princeton University Press, Princeton, N.J.

Gradmann, Christoph. 1996. Money, microbes, and more: Robert Koch, tuberculin, and the foundation of the Institute for Infectious Diseases in Berlin, 1891. Lecture given at conference, Pasteur, Germs, and the Bacteriological Laboratory, November 1996, Dibner Institute, Cambridge, Mass.

Harden, Victoria. 1992. Koch's postulates and the etiology of AIDS: an historical perspective. *History and Philosophy of the Life Sciences* 14:249–269. [See also the fascinating exchange of opinion between Peter Duesberg and William Blattner et al., *Science* 421:514–517, 1988.]

Mazumdar, Pauline. 1995. *Species and Specificity: an Interpretation of the History of Immunology*. Cambridge University Press, Cambridge, United Kingdom.

Mendelsohn, John Andrew. *Cultures of Bacteriology: Formation and Transformation of a Science in France and Germany*, in press. Princeton University Press, Princeton, N.J.

Weindling, Paul. 1992. Scientific elites and laboratory organization in *fin de siècle* Paris and Berlin: the Pasteur Institute and Robert Koch's Institute for Infectious Diseases compared, p. 170–188. *In* Andrew Cunningham and Perry Williams (ed.), *The Laboratory Revolution in Medicine*. Cambridge University Press, Cambridge, United Kingdom.

Preface

Robert Koch was one of the most important figures in medical science and was also the founder of bacteriology. Surprisingly, there has been no serious biography of Koch in English. Indeed, even the German-language biographies of Koch are dated and mostly inaccessible. The present book attempts to correct this deficiency.

This book began as an outgrowth of a larger project on the history of microbiology. When I began my research, I realized that I would have to spend considerable time on Koch because of his major importance. Yet there was no detailed summary of his work in English. There are many books in English about Louis Pasteur, the other leading figure of 19th century microbiology, but Robert Koch has attracted little attention in spite of his well-recognized importance. The biography by Dolman in the *Dictionary of Scientific Biography* is meritorious but is brief and hence serves primarily as a useful starting point. Most of the German-language biographies suffer from the sin of "hero worship". The biography by Bochalli is not only brief but rather stilted. The book by Heymann is rather too detailed and only deals with Koch up until his discovery of the tubercle bacillus. The book by Möllers, Koch's last assistant, is a useful source book of correspondence and dates but is ponderous and much too detailed for the general scientific reader. Both the Heymann and Möllers are out of print. The book by Genschorek, published in the German Democratic Republic, is also fairly brief. There are also a few fictionalized accounts of Koch in German, as well as a German-language movie, but nothing of any use to the student, scientist, or medical researcher.

The present book is based in the first instance on Koch's published work itself, and on the detailed Koch correspondence published by Heymann and Möllers, but has been greatly supplemented by my own reading in the bacteriological literature of the late 19th century. Probably at no time in the history of bacteriology did so much happen so quickly as between the year 1876, when Koch published his first work on the life cycle of the anthrax bacillus, and 1900, when Koch went into semi-retirement. In 1876, most physicians did not believe in the germ theory of disease, and medical practice was antiquated and based on incredible personal bias. By 1900, the Koch school of bacteriology was well established, and the disciplines of hygiene and public health had been placed on firm footings. Guided by *Koch's postulates*, investigators uncovered much of what is now known of the causes of the most important infectious diseases of humans and lower animals. Even today, Koch's postulates are considered in detail whenever a new infectious disease (such as AIDS), arises. To a real extent, we owe our current good health and longevity to discoveries made by Robert Koch and his school.

Koch was one of the true scientific revolutionaries. Beginning as a simple country doctor, he ended up his career as a Noble Prize winner and a dominant figure in 19th century and early 20th century medical research. His story can serve as an example of how a lone doctor, living and working in scientific isolation, can rise above his environment and become a major medical and public figure.

The manuscript for this book was read in its entirety by Professors Hanspeter Mochmann (Berlin-Buch), Werner Köhler (Jena), and Dieter Gröschel (Virginia), all of whom provided invaluable suggestions and corrections. Professor Mochmann also served as a gracious host during my visit to the various Koch sites in East Berlin. Dr. Masao Soekawa provided me with full access to the extensive Koch materials at the Kitasato Institute in Tokyo. Mr. Klaus Gerber of the Robert Koch Institute in West Berlin gave me full access to the Koch archives of that institute, and also provided me with photographs. Other photographs were provided by Dieter Gröschel, Hanspeter Mochmann, and by Dr. W. Presber of the Institute for Medical Microbiology in East Berlin. The Koch bibliography published here is based on a more extensive German-language version kindly prepared for me by Professors Mochmann and Köhler.

In preparing this book, I was fortunate in having access to an excellent library on the history of medicine at the University of Wisconsin-Madison. Librarian Dorothy Whitcomb provided endless amounts of help. Her courtesy in giving me long-term access to some of the most important Koch sources is greatly appreciated. Professor William Coleman gave me important advice and insights as this work progressed.

Jon Bartells did an excellent job as research assistant, digging out from the various library catacombs an amazing amount of valuable material. Kathie Brock copyedited the manuscript and provided many valuable comments. The Graduate School of the University of Wisconsin-Madison provided a modest amount of financial support. I would also like to acknowledge the good grace of my colleagues in the Department of Bacteriology of the University of Wisconsin-Madison for tolerating my decision to do work on the history of science. I hope I have justified their faith.

1

Introduction

From humble beginnings, and after an ordinary career as a country doctor, Robert Koch rose to the pinnacle of scientific achievement. Along the way, he established, virtually single-handedly, the new field of bacteriology. Koch's story is one of the most stirring in modern science and medicine. His concepts, methods, and discoveries pervade research in bacteriology and microbiology today as fundamentally as they have during the past 100 years. The human life span is now almost 20 years longer than it was when Koch began his work, and at least some of that increased longevity can be attributed to Koch's contributions. Whereas in Koch's time, the major killers of humans were infectious diseases, today infectious diseases are almost never causes of death in developed countries. At one time, tuberculosis alone was responsible for one-seventh of all deaths in Europe; today tuberculosis deaths are barely reported.

Looking back on medical research in the 19th century, it is conventional to pinpoint three giants in the field of infectious disease: Pasteur, Lister, and Koch. The work of Louis Pasteur (1822–1895) on fermentation and spontaneous generation provided the firm foundation needed for the development of microbiology as a science. Trained as a chemist, Pasteur in his later years embraced the study of infectious diseases with fervor, and was responsible for some of the most important steps toward the development of the field of immunology. Joseph Lister (1827–1912) was the founder of antiseptic surgery and is responsible more than anyone for bringing the surgeon out of the dark ages. And Robert Koch (1843–1910) was the founder of the field of medical bacteriology.

Although Koch and Pasteur fought bitter battles, their work was so complementary that one cannot imagine the one without the other. Whereas Pasteur built his work on a broad philosophical foundation, Koch's work was motivated strictly by medical questions. The work of both Pasteur and Koch had far-reaching impact on basic biological science, but they both are most known for their practical contributions. Louis Pasteur's primary medical contribution was in immunization, the development of procedures that could be used to protect or treat the individual. Robert Koch, on the other hand, made his major practical contributions in hygiene and public health, advancing innovations that dealt not with the individual but with the populace.

Koch's main contributions

Koch's contributions are, of course, the main subject of this book, but it is useful to lay out Koch's major accomplishments here, as an overview to what will follow.

1. He placed the **germ theory of disease** on a firm experimental footing.

2. **Koch's postulates** provide the essential experimental basis for any study of an infectious disease, whether in human, animal, or plant. Koch's postulates also apply in the broader field of microbial ecology of which medical microbiology is a part. The postulates thus apply to virtually any microbial process carried out in nature by a microorganism. Although these postulates were not completely new with Koch, it was Koch's experimental work that emphasized their importance.

3. Koch's **plate technique** for obtaining pure cultures was not only the foundation of bacteriological research, but provided the tool needed in studies on the genetics of bacteria.

4. Koch perfected the techniques needed to observe bacteria mic-

roscopically in diseased tissues. His work is fundamental to the field of **microscopic pathology**.

5. Koch made the first **photomicrographs of bacteria**, introducing this important tool of scientific communication to the microbiological research community.

6. Koch's **slide technique** is still the basis of routine laboratory study of bacteria.

7. He worked out the **life cycle of the anthrax bacillus**, showed the importance of **endospores**, and related this work to effective control of the disease. Koch's work on anthrax provided for the first time a complete life cycle of an infectious disease.

8. The contribution which gave him the most fame was his **discovery of the tubercle bacillus**. This discovery not only established without doubt the validity of the germ theory of disease, but placed on a firm basis the understanding of this most dreaded disease of humankind.

9. He isolated the bacterium which causes **cholera**, a discovery ranked by his contemporaries as important as or even more important than his discovery of the tubercle bacillus.

10. He perfected the essential procedures and methodology for **disinfection and sterilization**, thus making possible not only laboratory research but quarantine and other public health measures.

11. He discovered the phenomenon of **tuberculin sensitivity**, a discovery that provided one of the foundations for the discipline of cellular immunology. Also, Koch's idea that tuberculin might have therapeutic value, although wrong, was one of the first suggestions that specific therapies might be developed for individual infectious diseases.

12. Koch showed by careful epidemiological studies the importance of **water filtration** in the control of cholera, and by implication typhoid fever. This work led to the introduction of water filtration methods in large urban water supplies and resulted in major decreases in morbidity and mortality from intestinal infections. Water filtration has probably saved more lives than immunization and chemotherapy combined.

13. Koch was the first to discern that healthy human beings could be **carriers** of living pathogens and hence be responsible for

the spread of infectious disease, an important public health concept.

14. He participated in numerous national and international conferences on **public health and hygiene** and played a major role in the development of national and international regulations for quarantine and the control of infectious disease.

15. His work on the **etiology of wound infections** not only placed Lister's work on a firm scientific footing, but provided important insights into the whole question of bacterial growth in natural environments.

16. Koch was the founder of an important **school of bacteriology**, one of the most important schools of the late 19th and early 20th century. Some of the major figures who studied with Koch or in his laboratory were Emil von Behring, Paul Ehrlich, Richard Pfeiffer, Friederich Loeffler, Georg Gaffky, August von Wasserman, Shibasaburo Kitasato, and Carl Fränkel. Koch also had a strong impact on American bacteriology through his influence on William Henry Welch, Hermann Biggs, and T. Mitchell Prudden.

17. In his later years, Koch made major contributions to **tropical medicine**. Although much of this work was derivative of others, Koch was a strong promoter of quinine for the treatment of malaria and atoxyl for the treatment of sleeping sickness.

This lengthy list of contributions is justification enough for a biographical treatment of Robert Koch. But when we realize that the "Koch School" was actually the forerunner of much of what is now called *bacteriology* or *microbiology*, we can truly appreciate the magnitude of his contributions.

It would be unfair to either the man or the science to ignore another side of Koch's life. Robert Koch began his scientific career as an eager amateur (Figure 1.1), stealing a few hours away from his medical practice in order to experiment with bacteria and disease. He ended his career as an imperious and authoritarian father figure whose influence on bacteriology and medicine was so strong as to be downright dangerous. The story of this transition tells us something not only about Koch's personality, but also about the origins of the cult of personality in research. Early in his career, no one would have predicted that this hard-working and eager young researcher would turn into a crusty and opinionated tyrant.

(a)

(b)

Figure 1.1 *Two faces of Robert Koch. (a) As a young country doctor, 28 years old. (b) As a major figure in world medicine, 64 years old.*

2

Koch's Early Years

If I now consider the academic subjects which influenced my scientific development and especially my relationship to bacteriology, I must first indicate that I received from the University no direct influence for my scientific direction, because bacteriology did not exist in the University. Yet I would like to give special thanks to several of my teachers: the anatomist Henle, the clinician Hasse, and especially the physiologist Meissner, who awoke in me a feeling for scientific research.

—ROBERT KOCH[1]

He was born on December 11, 1843 in Clausthal, a small mining city in the Harz Mountains of Lower Saxony (Niedersachsen) (Figure 2.1). He was christened Heinrich Herrmann Robert Koch, although he was always known as Robert. Robert's father, Herrmann Koch, was a mining adminstrator who eventually became head of the mine. Robert's mother, Mathilde Juliette Henriette Biewend, was actually her husband's grand-niece.

Koch was the third in a family of thirteen, of whom eleven lived to adulthood. Although he was much closer to his mother than to his father, of greater influence than his mother was probably his uncle, Eduard Biewend, Koch's mother's brother. Eduard Biewend was devoted to nature study and photography and encouraged Robert in these pursuits. Thus Koch, the man who published the first photomicrographs of bacteria, was exposed early to the field of photography. Those were the heroic days of photography, when one prepared one's own photographic plates, and when it was a major accomplishment to obtain an image at all, let alone one of quality (see Chapter 7). Uncle Eduard also took Robert on short excursions into the countryside, where the goal

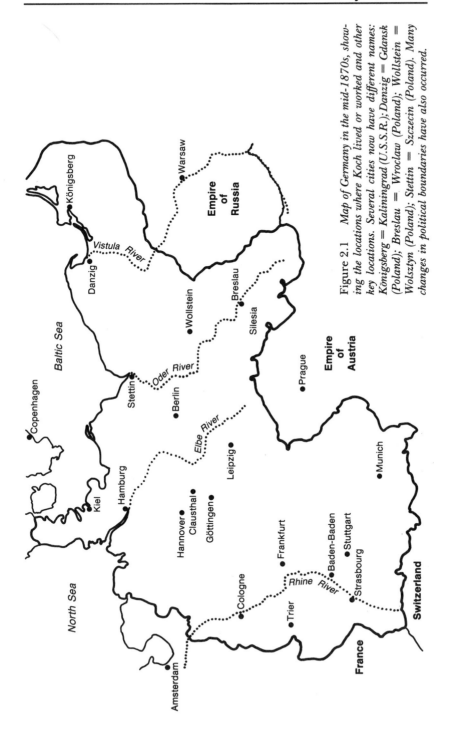

Figure 2.1   *Map of Germany in the mid-1870s, showing the locations where Koch lived or worked and other key locations. Several cities now have different names: Königsberg = Kaliningrad (U.S.S.R.); Danzig = Gdansk (Poland); Breslau = Wroclaw (Poland); Wollstein = Wolsztyn (Poland); Stettin = Szczecin (Poland). Many changes in political boundaries have also occurred.*

was nature study. Koch, attracted to animals at an early age, was an enthusiastic collector and classifier of beetles, butterflies, caterpillars, plants, and rocks.

A precocious youngster, Koch was able to spell words out of the newspaper at the age of four, and he started school at the age of five. He finished primary school and started in the Gymnasium in 1851 at the young age of eight. Koch was an industrious student, with a special aptitude for mathematics and the natural sciences. However, he showed little interest, and did poorly, in the classical languages, although he studied French and English enthusiastically.[2] Koch's fluency with English would be of great benefit to his field research during his trips to Africa, India, and the United States (see Chapters 15, 20, and 21).

With nine boys and two girls in the family, it was fortunate that Robert's father had a position of high responsibility so that as the family grew it could move to a large house (Figure 2.2). In addition to the large Koch family, the house was also home for two elderly aunts, a cousin, and a number of servants: a total of nineteen people! Of the eleven Koch children, Robert was the special favorite of his mother. Maternal influences on great scientists are often strong and Robert was apparently a real mother's joy. In addition to his passion for wildlife, he loved domestic plants and animals and took great pride, even as a nine-year old, in decorating the house with flowers.

At Easter 1862, at the age of 19, Robert presented his Abitur thesis at the Gymnasium (Figure 2.3). It was a short theme with the title: "How Odysseus Conquered Ajax". We have a report of his grades: Mathematics was "very good"; German, Physics, History, and Geography were "good"; and Religion, Latin, Greek, Hebrew, and French were only "satisfactory".[3] Such grades certainly do not presage a future Nobel Prize winner!

After Gymnasium, what was Robert to do? His older brothers Adolf and Wilhelm had emigrated to the United States and gone into business (see Chapter 21). Robert's father would have had him do the same, although his mother was reluctant to see her "favorite son" leave home. However, Robert apparently turned down the opportunity to emigrate and thus save his father money because of his fondness for a girl, Emmy Fraatz, who would later become his wife. Koch desired a university education and his school principal recommended that Robert study for a career in medicine or natural science, or perhaps for a career as a Gymnasium teacher. His principal apparently had some doubt about Robert's ability to apply himself, however, as he emphasized that Koch's

(a)

(b)

(c)

Figure 2.2   *Koch's early years. (a) The Koch family. Robert—leaning on his mother—was about 12 years old. (b) The home in Clausthal where Robert Koch was born. The house, now housing in part a flower shop, still stands at #13 Osteröder Strasse (corner of Bartelstrasse). (c) The large house that the Koch family moved to after the first house became too small. This impressive house, still extant, is in a favored location right in the main center (Kronenplatz) of Clausthal. (b) and (c) photographed by the author in 1987.*

(a)

(b)

Figure 2.3   *Koch's youthful years. (a) Robert Koch in 1861, just before he completed the Gymnasium. (b) A school fraternity "Concordia" in Clausthal in 1861. Robert Koch, 18 years old, is on the right in the rear (holding a blanket).*

success would depend upon his ability to concentrate his energies on a chosen field of study.

Vacillating from one possibility to another, Robert strongly favored a career in natural science, since this would have permitted him to partake of research expeditions to foreign lands. But the money for such an ambitious undertaking was not available, so Robert leaned toward one of the professions which had natural science as a base, such as medicine or teaching. Since the course of instruction for a teacher was shorter than that for a physician, he chose the former.

Koch entered the university at Göttingen at the age of 19. Göttingen, the most famous university in Niedersachsen, had many famous professors and was a fortunate choice for the young Robert. Among those at Göttingen was Friedrich Wöhler (1800–1882), the chemist who had synthesized urea and had thereby struck the first blow to the doctrine of vitalism; Friederich Gauss (1777–1855), the most famous mathematician of his time; Jakob Henle (1809–1885), the pathologist who wrote early and presciently about the germ theory of disease (see Chapter 5)[4]; and Rudolf Hermann Lotze (1817–1881), an instructor of medicine and philosophy who was one of the most outspoken opponents of vitalism.

In his first semester Koch studied trigonometry, solid geometry, physics, and botany, but soon decided that he did not want to be a teacher, so in his second semester he turned to the study of medicine (Figure 2.4). He then took anatomy from Henle, chemistry from Wöhler, and psychology from Lotze. Most great scientists have had behind them great teachers. Among the teachers Koch especially remembered[5] was the physiologist Georg Meissner (1829–1905), a master of animal experimentation. As we shall see, animal experimentation became the central feature of Koch's research program.[6]

Koch's main accomplishment at Göttingen was the winning of a prize of 30 ducats for carrying out a difficult anatomical study of the disposition of the nerves in the ganglia of the uterus. This study, for which ten months time was permitted, required exceedingly painstaking work, which Koch carried out under the direction of Wilhelm Krause, the director of the Pathological-Anatomical Institute. At the completion of this study, Koch presented his work, well illustrated with original drawings, under the title: "On the Presence of Ganglion Cells on the Nerves of the Uterus".[7] The report, dedicated to his father, was inscribed with Koch's motto "Nunquam otiosus" (*Never idle*). Koch wrote jokingly to his father: "Even a blind pigeon can find grain."[8] The 30 ducats of the Royal prize was the first money Robert Koch had ever obtained on his own. Koch's father was overjoyed with his son's success.

It is characteristic that Koch used his prize money to take a trip to Hannover to attend the 49th Congress of the *Gesellschaft Deutscher Naturforscher und Ärzte*. In addition to the opportunity to attend the lectures of the *Gesellschaft*, the meeting gave Koch the chance to see in person the renowned Rudolf Virchow (1821–1902), the most famous physician in Germany.

In the summer of 1865, Robert began a new experimental study

Figure 2.4   *Robert Koch as a university student, about 21 years old.*

under the physiologist Meissner on the formation of succinic acid in humans. These were the days before tissue slices, when biochemistry was done heroically in the whole animal. Koch, using himself as an experimental animal, carried out impressive dietary experiments on his own body, carefully analyzing his urine for succinic acid. (In one experiment, he ate a half pound of butter a day for several days!) Koch proved that with certain diets the body excreted large amounts of succinic acid. The study was published in the *Zeitschrift für rationelle Medizin*, a journal founded by Henle and Karl von Pfeufer (1806–1869).[9]

These youthful studies, so remote from his main opus of later years, already reveal the energy, industry, and scientific acumen with which Koch worked. As a fresh student of 22, he had already produced two substantial papers!

In addition to his independent research, Koch took numerous practical courses over the four years he attended the university. In addition, he served the various clinical services in the hospital, including surgery, obstetrics, psychiatry, and pathology. In January, 1866, at age 23, he passed his "Fakultäts-Examen", using the prize paper as his dissertation. In the same month he passed the "Examen rigorosum" and delivered, in Latin(!) an oral address on his succinic acid work: "De acido succinio in organismo humano", thus fulfilling his Doctor's degree.

For some scientists, the university degree leads immediately into a career in research, but for Robert, the way to research was to be long and arduous. Koch's behavior at this time gave no hint of future greatness. At the young age of 23, Koch certainly did not seem to be "driven" to research, despite the important accomplishments in his dissertation. It was only much later that he thrust himself with full force into a research career. Now, his graduation forced him to *decide* what he should do for a living. And, as we shall see, a difficult time of deciding he had indeed.

3

The Young Doctor and Husband

It's going awful here. We have to skimp and save every penny and I'm still not sure we'll make it. I keep telling Robert we have to leave here, that he must find a better position, but Robert has given up all hope . . . He can't make up his mind about anything without talking to his Papa.

—EMMY KOCH[1]

His medical studies completed, Koch traveled to Berlin to see the sights of the big city and to attend the lectures of Rudolf Virchow. Virchow, the most famous German physician of his day, was to become Koch's sometime opponent 20 years later (see Chapters 4, 8, 15, and 18). Koch visited the large hospital called Charité (Figure 3.1), which had 4000 beds and would be the scene of some of his later triumphs (and sorrows, see Chapter 18). Koch was appalled at the enormous size of the hospital, its huge lecture halls, and its bustling clinics. Nevertheless, he stayed three months, attending a course of lectures by Virchow.

On 12 March 1866, Koch passed the state medical examination in Hannover with no difficulty. Now that he was "licensed" to practice medicine, what should he do? He considered seriously becoming a ship's doctor, which would have satisfied his lust to see the world. But he finally gave up this "frivolous" idea in order to become engaged to his childhood sweetheart, Emmy Fraatz.

Emmy Adolfine Josefine Fraatz was the youngest daughter of an important official of the Evangelical Church in Clausthal. She and Robert had "kept company" before he had gone to Göttingen. However,

Figure 3.1 *The Charité Hospital as it looked at the beginning of the nineteenth century. The first part of this hospital was built in the 18th century.*

the engagement of Robert and Emmy came as a complete surprise to his parents. Robert's mother was delighted, as it meant that her favorite son would not be travelling to distant places.

However, if he were to marry, Koch must have a source of income. No longer could he count on his father for support. Koch thus took a position as a medical assistant at the Hamburg General Hospital, his first chance at real medical practice. In Hamburg, Koch began to show an interest in research, carrying out careful microscopical examinations of pathological materials. He also became acquainted at first hand with cholera, an infectious disease he would later study thoroughly (Chapter 15). The knowledge of the clinical picture of cholera which Koch gained in Hamburg would be of great use to him in his studies in Egypt and India 17 years later.

However, the position in Hamburg did not provide a suitable income for a prospective husband, so Koch began to apply for other positions. Even before obtaining a new position, he resigned and went home to await the results of his applications and to be with Emmy Fraatz. In October of 1866 he was offered, and accepted, a position as assistant in an institution for the education and care of retarded children, located in the tiny village of Langenhagen. This idyllic community, with a population of only 3000, was situated near Hannover. In this position, Koch would not only gain valuable experience, but he would be able to open up, on the side, his own private medical practice, thus making it possible for him to supplement his income. He quickly became a popular doctor with the villagers and his medical practice flourished.

Now that his private practice was prospering, his most passionate wish could be granted: he and Emmy could be married. The wedding took place on 16 July 1867 in the church in Clausthal; virtually the whole town turned out for the wedding. Robert was 24 years old (Figure 3.2). There could be no honeymoon, as Robert had to return immediately to Langenhagen to take care of a sick patient.

When he had first moved to Langenhagen, Robert had lived in the building of the institution, but now he found a seven-room apartment in the large house of a farmer (Figure 3.3a). His private practice continued to grow and he was soon able to purchase a "doctor's buggy", which made it possible for him to reach a wider patient population. Not only did the buggy help him to broaden his market, but it elevated his esteem in the eyes of his patients.

However, Koch did not restrict himself to his institutional duties and his private practice. As in Hamburg, he carried out microscopical stud-

(a)   (b)

Figure 3.2   *The young husband and father. (a) Robert Koch as a young doctor, about 24 years old. (b) Emmy Koch at the time of her marriage, 1867.*

(b)

(a)

Figure 3.3　*The peripatetic doctor. (a) The Koch residence in Langenhagen, near Hannover, where Robert and Emmy Koch lived from 1866–1868. (b) The Koch residence was on the upper floor of this building in Niemegk, near Berlin, 1868–1869.*

ies, not only in medicine but in the whole field of natural history. In the countryside were ponds and marshes, and Koch spent much time examining plant, water, and marsh samples microscopically for algae and protozoa. His wife, Emmy, was a great help in the collection of samples.

However, the idyllic life at Langenhagen ended in less than two years. Budget problems at the institution required that the position of Institute Director be merged with that of institute physician. At Koch's young age, 25, it was out of the question that he should be Director; he was thus out of a job.

Robert and Emmy now embarked on a period of uncertainty and transition while Koch tried one position after another, looking for something that would really be suitable for his life and goals. Things were made more complicated by the fact that Emmy was now pregnant. In order to spare her the ordeal of travel and uncertainty, Robert took her home to Clausthal.

At first, he opened a medical practice at Braetz, a small city between Frankfurt an der Oder and Posen, in the very eastern part of greater Germany. He moved to Braetz in June 1868 but soon found that the position was unsatisfactory, as he was unable to build a sufficient practice in competition with a well-established older physician. He quickly gave up and at the end of August, after less than three months, he moved to Niemegk (Figure 3.3*b*), a small village near Potsdam. His wife, still at Clausthal, gave birth on 6 September 1868 to a daughter, Gertrud, Robert's only child (Figure 3.4). Gertrud became Koch's pride and joy and she forged a bond between Robert and Emmy which kept them together through years of an increasingly unhappy marriage (see Chapter 19).

Emmy and Gertrud moved to Niemegk and there were three mouths to feed. However, the residents of Niemegk preferred faith healers to physicians and Robert's medical practice did not prosper. Living from hand to mouth, Robert decided to become a *Kreisphysikus*, or *District Health Officer*, a government position which provided a small but steady income and also permitted private practice. He applied to take the written examination for *Kreisphysikus*, but was not actually able to take the examination until later, after he had actually left Niemegk.

Things were not going well at all. One can see the first signs of dissension in the following letter which Emmy wrote to her father:

It's going awful here. We have to skimp and save every penny and I'm

Figure 3.4   *Gertrud (age 2) and Emmy Koch.*

still not sure we'll make it. I keep telling Robert we have to leave here, that he *must* find a better position, but Robert has given up all hope and now has the idea again to go overseas. He can't make up his mind about anything without talking to his Papa.[2]

Finally, in July 1869, after 10 months of struggle at Niemegk, Koch found a better location for his medical practice in Rakwitz (District of Bomst, in the Province of Posen, now in Poland but at that time part of Germany, see map, Figure 2.1). Rakwitz was no larger than Niemegk but it had an extensive hinterland from which patients could also be drawn. The Kochs lived in a large two-story house which also held the post office. The region had a mixture of German and Polish inhabitants, and Koch hired a bilingual servant girl who also served as his interpreter. At first, his practice was modest, but it increased notably after he succesfully treated a large estate owner, the Baron von Unruhe Bomst, who had accidentally wounded himself with a revolver.

Koch's reputation for industriousness and medical skill soon spread and he found patients coming from many of the surrounding villages as well as from the countryside. The move to Rakwitz was especially important for Koch's subsequent development as a medical scientist, because it brought him near Breslau, the city where he received important early recognition (Chapter 6).

However, although the medical practice was good, the work was hard. In December 1869 he wrote his father:

> Many days I don't have a second to myself. On my birthday I had to make five separate trips to the country and I was on the go from 4:30 AM until 11:30 PM. However, on some days I have quite a few free hours, which I can then use to follow my studies.[3]

In Rakwitz, Koch turned again to natural history. He kept a variety of animals on the grounds: pigeons, dogs, chickens, cats, and even, for a while, foxes. He also took up beekeeping. He was quite a bon vivant, visiting the local restaurants, bowling establishments, and *Bierstube*. In his spare time he played the zither. To his patients, he soon became the beloved "Doctor", and he also became the drinking friend and honored guest of the mayor and the pharmacist.

The pleasant existence in Rakwitz came to an end with the outbreak of the Franco-German War of 1870. Koch was exempt from military service because he was nearsighted, but he volunteered anyway as a physician in a battlefield hospital. He served in Neufchateau and Orleans in France, where he became acquainted with typhoid fever and battle wounds. His war experiences must have been invaluable for him during his important research on wound infections eight years later (see Chapter 8). His three younger brothers, Hugo, Albert, and Ernst, were also in the war and Robert saw them often, taking a "fatherly" interest in their welfare. According to his letters to his father, he learned more during his brief military service than he would have learned in a half year in a surgical clinic.

However, his hometown of Rakwitz needed its popular doctor back, and he was called to return even before the war ended. At the end of January 1871, Koch went home to Clausthal to see his sick mother, who died shortly afterward, and then returned to Rakwitz, where he was received with acclamation. He had learned much during his brief war sojourn, but it was good to be back. Despite the predominance of Polish, Koch for the first time felt that he had established a "home". But at age 28, Koch was on an important new threshold.

4

Steps Toward Maturity: Koch in Wollstein

What a doctor! There was something special about him. If Dr. Koch came into your sick room, you immediately felt calm and secure.[1]

Despite his increasing fortunes at Rakwitz, Koch was not content to spend his whole life as a simple country doctor. Even before going to Rakwitz, he had applied to take the written examination for District Medical Officer (*Kreisphysikus*), and now the opportunity to take the examination arose. In January 1872 he took and passed the exam, which gave him the opportunity to apply for a vacancy if one should present itself. Now he received an invitation from one of his important patients, the Baron von Unruhe Bomst, to apply for the newly created position as *Kreisphysikus* in Wollstein. Koch applied in person, was accepted, and in April 1872 he moved to Wollstein to begin one of the most important periods of his life. When Koch arrived at Wollstein he was just a 29-year-old physician; when he left eight years later he was on his way to becoming a renowned world-class scientist.

Wollstein was a small city of only 3000 inhabitants, but it was situated on a beautiful lake within a pleasant forested district, just the location for a physician inclined toward natural history. Wollstein had been founded as a German city in 1458, even though it was now surrounded by mostly Polish countryside.[2]

The Koch family lived at 12 *Strasse am weissen Berge* (now called Dr. Robert Koch Street, *Ulica Dr. Roberta Kocha* in Polish) from 1872 until 1880.[3] The living quarters consisted of four large rooms and a kitchen on the upper floor of a two-story house. The house, with its high Gothic bay windows and its columnated balcony, stood out sharply against the one-story farmhouses on either side (Figure 4.1). The large room in the left rear served as Koch's clinic, the patients waiting on the landing outside until called. This examination room, which faced west and south, had large windows and was bright with sun and light. It was in this room that Koch carried out his pioneering researches on anthrax and the germ theory of disease. When Koch's research began to become ever more dominant, Emmy Koch divided the room into two parts with a curtain; behind this curtain Koch set up his first laboratory. At the window, he made an arrangement so that sunlight could be brought in for photomicrographs (Chapter 7) and it was here that the first photomicrographs of bacteria were taken. A crowded laboratory, but suit-

Figure 4.1 *Robert Koch's house in Wollstein, photographed in 1987. The large double doors lead to an open coachwagon entry area, and on through to the garden in the rear. The Kochs lived on the upper floor. In addition to the three front rooms, in the rear on the left was a large room (Koch's clinic and laboratory) and on the right the kitchen.*

able for one person. Only Emmy and Gertrud were permitted to enter the workroom. We are told[4] that when sunlight was being used for photomicroscopy, Emmy Koch would be stationed outside to watch for clouds that might spoil the exposure. Koch called his wife his "*Wolkenschieber*", his "cloud chaser".

By careful budgeting of family expenses, enough money was saved so that a good microscope could be purchased. It was a Hartnack, one of the best available at that time. However, purchasing the microscope meant that Koch had to sacrifice the purchase of a much-needed carriage for housecalls in the country. It is revealing that Koch started real research as soon as he had enough money to buy equipment.

Despite the modest surroundings, these were perhaps the richest and happiest years of Koch's life. His daughter was growing into girlhood (Figure 4.2) and he played with her delightedly, lifting her high in the air to the sound of excited squeals, imitating animal calls for her, and carrying her on his back. He looked carefully after her schooling, bought her toys and pets, or whatever else she wanted. He amused her by telling her stories of his own childhood and of the clever games that he and his brothers had played. What hidden emotions are suggested by the

Figure 4.2   *Gertrud Koch in Wollstein, at age seven.*

following letter, written by Robert to his daughter Trudy on her eighth birthday, when she was visiting her grandparents in Clausthal?

> Dearest Trudy,
>   I was so pleased to hear from Mama that you have been a good girl. I am so pleased! And the best of luck on your birthday. Eight years old already! Such a big girl! And now you must become a smart girl, study hard in school, help your mother in the kitchen, arrange the flowers, feed the animals, and help me by cleaning my microscope slides and collecting algae. But all these things you can already do! And each year you will be able to do more and more. Finally, Papa and Mama will be able to lounge around the whole day in our comfy chairs while our darling Trudy cooks for us and takes my notes and works with the microscope. Oh, won't that be a nice time? And now, don't stay away too long from your old father. The animals are looking all over for you and Julka [the Polish servant] keeps sighing for her Trudy. And each time I hear the door open, I am sure that it's my little girl coming. But instead, it's just a stranger. Come home soon to your loving Papa!"[5]

Koch's medical practice kept growing. He was extremely popular with his patients, and their loyalty was high. They gave him presents and wild game, and would travel long distances to be attended by him. Years later, after Koch had become famous, an elderly Wollsteiner remembered those days:

> What a doctor! There was something special about him. How often did I hear my mother say: "If Dr. Koch came into your sick room, you immediately felt calm and secure."[6]

As the most popular doctor in the area, he could have kept busy with nothing but his patients. But practicing in such a rural area had many difficulties. The Polish farmers often came for him late at night, since all day long they were working in their fields. On the other hand, there were special pleasures in driving out to see patients. He could then observe the natural beauty of the countryside and collect samples, and he could take his little daughter with him.

The garden of the house was a constant source of pleasure. Here, among the flowers and trees, Koch opened his daughter's eyes to the beauties of nature.[7] He loved to walk in the garden, taking pleasure in his long-stemmed pipe. In time, he filled the garden with a veritable menagerie of experimental animals: guinea pigs, rabbits, mice, even two monkeys, all of which his wife and daughter cared for.

This was a time when electricity was still a novelty. Might it have healthful effects? Koch made a machine (*Elektrisiermaschine*) which generated an electrical charge. He installed it in his examination room, where he used it to "treat" patients. One time he even "electrified" his daughter and the members of her play group, standing them all in a circle and laughing heartily at the cries of the girls as their hair stood on end! He even made a portable electrical machine that he could take with him on house calls, and tried to get a Berlin factory to make a commercial version, but nothing came of it.

What was Koch like in these days? His son-in-law and scientific collaborator Eduard Pfuhl has given us some insights, based on Koch's daughter's remembrances and on stories he had heard from Koch himself. Koch enjoyed his food and drink. He kept in his wine cellar barrels of the best Hungarian wines, both red and white. When he could finally afford it, he bought himself an elegant fur coat. From childhood, he wore nothing but smooth cotton stockings, rather than rough wool. He enjoyed reading humorous and comic writings. One time he dislocated a leg tripping on a carpet. He bandaged the leg himself and kept on working, despite the pain. One time late at night a fire broke out in a kerosene warehouse near his home. He was nearby with friends drinking beer and hurried home to calm his wife. His friends and his wife wanted to immediately start saving all their possessions but Koch refused: "We're fully insured. The most important thing is to save all the barrels of kerosene in the warehouse that haven't yet caught on fire." Then he and his friends rolled all of the kerosene barrels out into the open, thus not only saving the kerosene but confining the fire. While moving the barrels, Koch stepped on a nail and hurt his foot so badly that he was confined to the sofa the next day. He sent his patients to one of the other doctors, but some patients refused to see anyone but Dr. Koch, so he held "office hours" while flat on his back. Although Koch's pictures often show him as humorless and stern-faced, he actually had a refined sense of humor. Since he was nearsighted, he always wore glasses, and this made him look a little "remote". However, according to Pfuhl, he looked even sterner when he took his glasses off.[8]

In addition to his medical practice, Koch had all the duties of the *Kreisphysikus*. He had responsibility for smallpox inoculations, for writing death certificates, for giving advice on general public health matters, and for overseeing the local hospital. He also worked hard at keeping up with advances in medicine, reading many medical journals and books. As he read, he would make careful written notes.

But not only did he find time for all these medical matters and his research, he also took an interest in archaeological explorations. In the neighborhood of Wollstein were many prehistoric sites, with pottery pieces easily found in the sandy soil of the area. Koch brought his best "finds" home to show his wife and daughter. His hobby of archaeology led him to a meeting in Wollstein with the eminent Rudolf Virchow. At this time, Virchow was intensely interested in prehistory. The local farmers frequently found decayed bones of humans and animals and other ancient remnants as well as pottery. At Koch's suggestion, his friend the Baron sent a small collection of these findings to Virchow, in Virchow's role as head of the Berlin Gesellschaft für Anthropologie, Ethnologie, und Urgeschichte. Pleased with the finds, Virchow visited Wollstein on 1 May 1876, where he was shown around by the Baron and Koch himself. The small party went to a large estate in the area where a "dig" was already in progress. Koch, his wife, the Baron, and Virchow all donned protective clothing and descended into the diggings with lead trowels, uncovering new finds, to the accompaniment of much laughter and joking. After a few hours, the digging ended and the happy party was treated to a hearty breakfast in the manor house of the estate. Only two years later when Koch had completed and was demonstrating his important work on infectious disease, he received a frosty reception in Berlin from this same Virchow (Chapter 8). At the end of Virchow's visit, Koch and his friend the Baron were made members of the Berlin Gesellschaft, and the choicest items from the dig were exhibited in the Berlin Museum. Koch continued to maintain an interest in archaeology throughout his life and sent collections to Berlin from his many later expeditions around the world.

It is amazing that Koch found time, along with his official duties as *Kreisphysikus*, his extensive medical practice, and his private researches, to still follow a hobby of archaeology.

5

The Lone Scientist: The Work on Anthrax

Never, surely, could a man have found himself in a position less favourable for scientific research—poor, humble, unknown, isolated from sympathy and from the scientific appliances which are the necessary tools of the investigator. Yet he was a man of too strong a character to allow himself to be warped by the position in which he found himself, or to be diverted from the line of work which was most congenial to his nature.

—ARTHUR CONAN DOYLE[1]

When Robert Koch had finally built a lucrative medical practice, he started doing research. At first he began to study hygiene and public health problems related to mining and smelting, visiting his father in Clausthal and his brother Hugo, who was director of a mine and smelter in Tarnowitz. However, it was after he took a major trip through all the research centers of Germany that he began, in earnest, to study the role of bacteria in infectious disease.

He had begun to study the disease *anthrax* even before his big trip, but it did not become his passion until he returned. The purpose of the trip, taken in the fall of 1875, was to attend several scientific and medical meetings and to visit research centers. This trip represented a watershed for Koch. Seeing the scientific world as it was, he returned home refreshed and invigorated, and motivated to make significant scientific contributions of his own.

Koch's main activities on his trip were the meetings of the *Deutsche Gesellschaft für öffentliche Gesundheitspflege* in Munich, and the *Gesellschaft Deutscher Naturforscher und Ärtze* in Graz (Austria). During the trip, Koch also visited important research laboratories, most notably that of the famous hygienist Max von Pettenkofer in Munich. This trip was not only a scientific but a cultural trip, as Koch visited theaters, museums, and various scenic attractions. He also ate and drank well, describing in his diary of the trip each restaurant which he visited. (He was fascinated by the quaintness of the Hofbräuhaus, still a landmark in Munich, and commented on the good but cheap beer.) This trip provides the first suggestion of the wanderlust which was so evident in Koch's later life. When he returned home, he began the studies on anthrax which were to lead him to world fame.

Infectious disease before Koch

Although Koch's work put the germ theory of disease on a firm footing, Koch did not begin his work in a vacuum. Many workers had come before him, not the least Louis Pasteur himself. But Koch's first published work on anthrax was so perfect, so detailed, so complete, that it quieted all significant objections to the germ theory of disease. And it started Koch himself on the career that would carry him to the highest pinnacles of scientific society, culminating in the Nobel Prize itself.

Where was medical bacteriology when Koch began his work? Lister's procedures for antiseptic surgery, based on Pasteur's work on the microbial basis of fermentation and putrefaction, were well accepted, and this gave strong credence to the hypothesis for microbial origins of disease. But evidence on the role of *specific* microbes in *specific* diseases was lacking. Jacob Henle (1809–1885), one of Koch's teachers at Göttingen (Chapter 2), had written about the so-called *contagion* as a living entity:

> . . . consider the case of a thorn which has been thrust into a finger and causes inflammation . . . if the thorn is removed it can be stuck into the finger of another person, and cause the same disease a second time. It is not the *disease* which is being transferred, but the *cause of the disease.* Now suppose that the thorn could reproduce itself in a diseased body, or that every small part of the thorn could turn into a new thorn. Then, through the transfer of each small part of the thorn, one could cause the same disease in others. The thorn, not the disease, is the parasite. . . . The contagion, as we consider it, is similar to the thorn. The contagion is not itself the disease, but the *inducer* of the disease. . . . As an example, when a needle is immersed in a solution containing a small amount of smallpox vaccine, this is sufficient to cause an infection. This effect by such a small dose depends on the ability of the agent to reproduce itself. . . . This is therefore further proof that the contagion is alive.[2]

Henle's writings, based on careful observation of the communicability of disease, clearly distinguish between the parasite and the disease. But even more, Henle points the way toward an experimental analysis of contagious diseases:

> After it has been shown that the contagion is alive, there still remains the question of how the contagion works to bring about its damage. If it could be possible to prove that a contagion can be cultured outside the body . . . then such a contagion could only be a plant or animal . . .

The key point here, and one which was at the basis of all work after Henle, was that the contagion reproduces and might be reproduced outside the body. This raised the possibility of an experimental approach to the study of contagious disease. But Henle's work, just quoted, was published in 1840. What had happened by the 1870s to advance the field?

In 1875, the controversy over the parasitic nature of infectious diseases was occupying medical congresses throughout Europe. Some of the major pathologists of the day, such as Rudolf Virchow in Berlin and Theodor Billroth (1829–1894) in Vienna, denied the role of microorganisms in infectious disease. But an especially passionate advocate of the parasitic origin of disease was Edwin K. Klebs (1834–1913), a German physician who at this time was Professor of Pathological Anatomy at the University of Prague (then part of the Austro-Hungarian Empire).[3] A student of Virchow, Klebs later became one of Virchow's opponents.[4] Klebs was a passionate believer in Henle's teachings and carried out inoculation and culture studies for such conditions as diphtheria, smallpox, and wound infections. Klebs insisted:

> It is not the disease, but the cause of the disease, which reproduces.[5]

In fact, Klebs outlined the experimental steps that would later become known as *Koch's postulates*:

1. Careful microscopic study of the diseased organ.
2. Isolation and culture of the germ associated with the disease.
3. Production of the same disease by inoculation of this cultured germ into healthy animals.

However, although Klebs' ideas were good, his bacteriological technique was bad. He was unsuccessful in obtaining pure cultures, and he was thus unable to convince his colleagues of the validity of his ideas. As the Leipzig surgeon Karl Thiersch (1822–1895) said:

> My heart says "yes" to bacteria, but my reason says "wait, wait".

And at another place:

> Does disease follow bacteria, or do bacteria follow disease? We still don't know the answer.[6]

One of the problems was that at this time, the scientific study of bacteria was quite limited. We will discuss in Chapter 6 the important work of Ferdinand Cohn, the Breslau botanist, whose writings on bacteria were widely known. However, the most important forerunner of Koch was the French scientist Casimir Joseph Davaine (1812–1882), who showed clearly the transmissibility of the disease anthrax.[7] Davaine had been influenced by Pasteur's work on the microbial origin of fermentation published just a few years earlier (1857). It had been known since the work of Aloys Pollender (1800–1879) in 1849 that rod-shaped structures of bacterial size were present in the blood of cows that had died of anthrax.[8] But Davaine showed that if a small amount of blood was taken from a diseased animal and injected into a healthy animal, the latter succumbed to anthrax. Davaine also showed by inoculation experiments using fresh or dried blood that the rod-like structures present in the blood were probably living bacteria. He called these structures *Bacteridia*, a term subsequently used by Pasteur, but rarely by Koch. However, because it was known that anthrax could also be acquired from contaminated soil, it was not clear that the "contagion" was obligatorily connected with the disease. It is important to understand this latter point, which is a rather special complication of anthrax because the causal agent is a spore-forming bacterium. Up to this time, contagion implied *direct transmission from person to person*. If a disease could also be acquired from the soil, then it was not a contagious disease.

The advantages of anthrax as an experimental model were several:

1. The causal organism was very large and thus easy to see even with the microscopes available when Koch began his work.

2. In the later stages of the disease, the organism was found in large numbers in the blood, thus making sampling easy.

3. Inoculation experiments with animals could be readily performed, making it easy to establish the disease in an experimental system.

4. Anthrax was an important disease of domestic animals, and hence of great economic concern, but it also occurred occasionally in people, making it also a medical problem.

Koch's initial work on anthrax

Robert Koch began his work on anthrax in the same way he carried out microscopic studies of algae and protozoa, as a hobby. And although his initial work was motivated by a desire to understand a particular public health problem in his district, anthrax in the end provided a *model* for studying the bacterial nature of infectious disease. Koch studied anthrax not to find a cure, but as a means of studying an important basic problem.

In the district of Bomst, an outbreak of anthrax occurred that was causing extensive fatalities in sheep and cattle; occasionally humans were also infected. In 1873, Koch began his initial microscopical observations on blood from sheep that had died from anthrax. On 12 April 1874, Koch obtained his first observations that suggested that Davaine's *Bacteridia* might form spores:

> The bacteria swell up, become shinier, thicker, and much longer. Slight bends develop. Gradually a thick felt develops. Within the long cells, cross walls appear and small transparent points develop at regular intervals.[9]

These are the first recorded observations of anthrax endospores. Observations of these endospores spurred Koch on to heroic efforts. Sparing neither energy, time, nor money, he worked feverishly. His family—and indeed, everything else—came second to his research. Emmy Koch has written:

> Through careful saving of money, a better microscope was obtained. How happy Robert was! Then came another day of celebration when a better microtome was purchased. Then later, a spectroscopic apparatus. . . . Every physician knows how often patients come and disturb the doctor with only minor illnesses. It was my job to find out first how sick a patient really was, and to send away those who didn't really need medical attention. In that way, Robert could often remain for hours at his work.

In order to find more time, Koch managed to send some of his patients to a colleague, a Dr. Zielewski, who had settled in Wollstein. But Koch's work on anthrax only progressed in earnest after the trip in the fall of 1875 mentioned earlier in this chapter. His work was now bolstered by the acquisition of fresh, highly infectious material, which he came by accidentally in the hide of an animal which had died of anthrax. A police officer had confiscated the dead animal and Koch, as the *Kreisphysikus*, was called in to examine it. Turning to his microscope, Koch found that the blood associated with this hide teemed with bacteria.

On the 23rd of December 1875, a fateful day, Koch inoculated a rabbit through the ear and in the back with a tiny drop of blood from this infected hide. The rabbit rapidly became ill and died 24 hours after inoculation. In great haste, as it was Christmas Eve, Koch surgically removed the inoculated ear, excised the skin on the back at the site of inoculation, and placed both of these pieces of tissue into a jar of alcohol, after first checking to be certain that the *Bacteridia* were present ("in moderate numbers", Koch wrote in his note book).

The next morning, Christmas Day, with no demands from patients, Koch could study his dead rabbit at leisure. Carefully, he dissected the inguinal gland and the ear gland from the corpse and examined fluid from them microscopically. The samples were teeming with bacteria. A new rabbit was inoculated in both corneas with material from the inguinal gland.[10] Nothing happened, so on 29 December he made tiny surgical slits in the cornea of the left eye so that fluid was expressed. He then introduced a tiny bit of tissue from a bacteria-containing lymph gland. On 3 January 1876 the animal died and bacteria were present in the spleen, the cervical gland, and in the blood, as well as in the turbid fluid of the eye. The growth of bacteria in the corneal fluid gave Koch a brilliant idea: To use the aqueous humor of the rabbit's eye for the *artificial culture* of the anthrax *Bacteridia*.

How quickly Koch had moved from simple microscopic observations to animal experimentation to artificial culture! He was following Henle and Klebs, of course, but the critical thing turned out to be the selection of a suitable culture medium. Earlier attempts to culture the anthrax bacillus by Koch and others had failed, but the aqueous humor of the eye worked beautifully. Soon Koch was obtaining the aqueous humor of the cattle eye at the slaughter house and thus had a plentiful supply of culture medium.

Quickly, Koch discovered the importance of using a warm temperature, 30–35°C, to grow the bacterium, and the need for oxygen. He was using slide cultures now, which he had sealed with paraffin to keep from drying. However, he noted that preparations which he had not sealed, and which were almost dry, showed better bacterial growth than those that had been sealed, despite the drying. Deducing the importance of air, Koch manufactured small glass culture chambers which permitted good aeration but protected the cultures against drying.

Now came the major discovery. Examining some of his slide cultures, he observed long filaments containing refractile spheres. The spheres remained arranged in rows even after the filaments broke down and disappeared. Koch, the naturalist, immediately interpreted these refractile objects as resting spores (in German, *Dauersporen*).

We can be truly amazed by this work. In slightly more than one month, beginning on Christmas Eve, Koch had worked out the whole life cycle of the anthrax bacillus! Much work remained to be done, but the framework of future research was clear. In the aqueous humor of the eye, Koch had found a culture medium which permitted growth and spore formation. Within weeks, Koch had worked out in great detail the biology of the anthrax bacillus. Taking his lead from the classification of bacteria presented by Ferdinand Cohn (see Chapter 6), he used the species name which Cohn had used: *Bacillus anthracis*. The use of a species name was not just a convenience: it implied that anthrax was caused by a specific organism.

Closing the gap: the bacterial life cycle

The discovery of spores was critical. Davaine and others had shown that animals could obtain anthrax not only from other infected animals but from the soil. The knowledge of soil-borne anthrax had cast serious doubt on the validity of a living contagion. But if the bacterium formed "resting spores", then everything became clear. The resting spores could remain *alive* in the soil and serve as a later source of infection. In Koch's words:

> The bacteria form spores which posssess the property of growing into new bacteria after longer or shorter resting states. All my further experiments were directed to discovering this suspected developmental stage of the anthrax bacillus. After some fruitless experiments, it was possible finally to achieve this goal and therefore to determine the true etiology of anthrax . . .[11]

Koch's work benefitted greatly from his natural science background. The concept of a "life cycle" is familiar to one trained in natural science, but not necessarily to those whose training is restricted to medicine. As we have seen, Koch not only received early training in natural science, but continued to maintain an interest in nature after he began medical practice. The fact that he kept various kinds of animals as pets meant that he had experimental animals available for work when he needed them. Koch's ability to take the giant leap from the life cycle of the anthrax bacillus to the etiology of anthrax undoubtedly depended in part upon this natural history background. However, one trained *only* in natural history might not have been able to do the expert pathological work which Koch had done. It is not accidental that the title of Koch's first paper on this subject is: "The Etiology of Anthrax, Based on the Life Cycle of *Bacillus Anthracis*".

The relationship between "etiology" and "life cycle" is one that few others besides Koch could have made!

Imagine the difficulties Koch experienced in his research. For his culture work, he needed a proper incubation temperature. There was no household electricity in 1876. A kerosene heater had to be used, yet how could the temperature of such a burner be regulated? Ingeniously, Koch filled flat dishes with moist sand, laid filter paper pieces on top, and placed his slide cultures on top of the filter papers. He then heated the sand with a small kerosene flame, adjusting the height of the flame to obtain in the sand the desired temperature. "To all who would undertake such experiments without gas or a regulator, I recommend this method highly. It is necessary to fill the kerosene reservoir only once a day, in order to have a temperature which varies no more than 1 to 2 degrees."

However, Koch was not content to infer the life cycle of the bacterium. He wanted to observe it directly:

> Since I now knew the conditions under which I could obtain satisfactory growth of Bacillus anthracis . . . I began to apply these conditions to microscope slide cultures, so that I could observe directly the changes in morphology of the bacilli.

Koch used a heatable stage which he purchased from the company of M. Schulze:

> The microscope must be placed on a support in order to bring the chimney of the kerosene lamp under the arm of the heatable stage. A single small

flame, standing under the center of one arm, maintained the stage for days at the necessary temperature.[12]

However, up to this point we have discussed only a part of Koch's pathfinding research. Koch did not confine himself just to culture studies. And here is where his medical training took over.

> It has not been possible for me to observe the multiplication of the bacteria directly in the animal. But multiplication can be inferred from the inoculation experiments which are described below. I have used the mouse as my experimental animal, as it is simple to use. . . . In most experiments I inoculated the mice at the base of the tail, where the skin is loose and covered with long hair. . . . I have made a large number of inoculations in this way, using fresh anthrax material, and in every case I have had a positive result, and I believe therefore that the success of the inoculation can be used as an indication of the viability of the bacilli inoculated. . . . Partly in order to always have available fresh material, and partly to discover if the bacilli would change into another form after a certain number of generations, I inoculated a number of mice in series, one after the other, each time using material from the spleen of a mouse which had just died. The longest series of mice treated in this way was twenty . . . In all animals the results were the same. The spleen was markedly swollen and contained a large number of transparent rods which were very similar in appearance and were immotile and without spores. . . . A small number of bacilli could always develop into a significant mass of individuals of the same type . . .[13]

Although Koch was not the first to use experimental animals for the study of infectious disease, he was the first to use them in an integrated way with other experimental methods. And his subsequent work on infectious disease depended to a major extent on animal work. In Koch's day, there were no commercially available experimental animals such as mice. One used gray house mice, which had to be caught in the horse barn.[14] Because these mice were wild, they were hard to work with. In order to get a mouse out of the tall glass jar that it was kept in, Koch used an old bullet extractor (*Kugelzange*) which he found in the old case of instruments that he had used during the Franco-German War. Later, such forceps became widely used in medical bacteriology and became known as "Koch's mouse forceps". White mice became available to Koch only after his anthrax work was finished.[15] Emmy Koch fed and cared for the infected mice. She also incinerated the dead and autopsied mice in the large oven (*Kachelofen*) present in Koch's patient examination room.

The etiology of anthrax

Having studied the bacillus, and having carried out careful inoculation experiments, Koch then made the intellectual leap to the etiology of the disease itself. He carefully pointed out that his animal experiments had only been done with rodents, whereas the disease in domestic animals was in ruminants; nevertheless, he drew important public health conclusions from his work.

> We thus see that anthrax tissues, regardless of whether they are relatively fresh, putrefying, dried, or years old, can produce anthrax when these substances contain bacilli capable of developing spores of Bacillus anthracis. Thus, all doubts regarding Bacillus anthracis as the actual cause of the disease must be dispelled. Bacillus anthracis is indeed the contagion of anthrax. The transmission of the disease in fresh blood occurs rarely in nature, only in persons who come in contact with blood or tissue juices while killing, cutting, and skinning animals infected with anthrax. . . . But, the great percentage of infections are produced only by the penetration of the spores of Bacillus anthracis into the animal body. For the spores can survive in an amazing manner for many years. When these spores have once formed in the soil of a region, there is good reason to believe that anthrax will remain in this region for many years. . . . A single cadaver, handled improperly, can furnish almost innumerable spores . . . If it would be possible to discover how the spores of Bacillus anthracis were disseminated, and the conditions under which the contagion renews itself *de novo*, it might then be possible to prevent the growth of the bacillus and therefore reduce the incidence of disease or perhaps even exterminate it entirely. . . . The present hygienic measures against anthrax are limited to notification of the authorities, the burial of cadavers, disinfection, and quarantine of the town affected by the plague. [However, these measures are of little value if the spores can remain alive for many years in the soil.] We must, therefore, seek other measures in order to free the herds from this destroyer and to protect thousands of persons from an agonizing death.[16]

An amazing piece of work, and one that was to cause great excitement in the medical community. But Koch, working in isolation in the small village of Wollstein, was full of self-doubts. Was he correct? Had he made a mistake? Was Bacillus anthracis really the cause of anthrax? Spores such as he observed had never been described before. Were they really spores? And were they really important in the life cycle and etiology of the disease? Should he publish his work; expose it to scientific scrutiny? Was he, a lone doctor using primitive techniques, wrong? "I wanted to continue my studies for longer before publishing them, but just at that time a well-known botanist declared that the rods associated with anthrax were crystalloids, so it seemed to me that I should publish my observations."[17]

Wracked with doubts, Koch came to the conclusion that he should demonstrate his most important observations on the life cycle of the anthrax bacillus to the best possible authority on the biology of bacteria: Professor Ferdinand Cohn of the University of Breslau. Koch's decision to write Cohn brought to an end a lengthy period of doubt and self-criticism.

And when Koch finally did make his work known to the scientific world, it was received with acclamation.

Koch was 32 years old when his work on anthrax was published. What drove Koch to a study of infectious disease? How did this man, who had been content to use his microscope to observe algae and protozoa, and who had dreamed of sitting back in his easy chair and letting his daughter wait on him, turn to a study of bacteria as sources of contagion? The true scientist, Koch was driven by curiosity—by a desire to know. Starting out with a relatively simple problem, he followed the experiments where they led him. As one result occurred, it suggested another experiment; and another; and another. A research scientist understands how this happens. You follow the trail where it leads until, one day, you realize that you are in a new country. That what you have is more than a series of results, but the basis of a *new concept*. Once the concept is formulated, new experiments are suggested, and now one is really on the way. Koch's turning point came that night when he saw spores in a slide culture and related them to the etiology of the disease. From then on, there was no turning back. Koch moved inexorably forward, forging the bacteriological revolution that was to have such major impact on medicine and human society.

6

First Recognition: Koch and Cohn

How happy I was when Robert finally wrote to Professor Cohn and went to visit him.

—Emmy Koch[1]

As discussed in the last chapter, Koch's work on anthrax was carried out in complete isolation from the scientific world. The work was brilliantly conceived and executed, and it was to make Koch's name known throughout the world. But when he began to think about publishing it, Koch was wracked with self-doubt. Was the work any good? Had he made some horrible blunder? Before he published it, he needed to discuss it with a knowledgeable scientist. Fortuitously, Professor Ferdinand Cohn (1828–1898), one of the foremost botanists in the world and the author of an important book on bacteria, lived in the city of Breslau, not many hours by train from Wollstein. Koch determined to demonstrate his results to Cohn.

Ferdinand Cohn

Before discussing the historic Koch-Cohn meeting, we should briefly discuss Cohn himself. Ferdinand Julius Cohn (Figure 6.1), the eldest of three brothers, was born of Jewish parents in Breslau[2] on January

Figure 6.1   *Ferdinand Cohn.*

24, 1828.[3] Cohn was a brilliant child who excelled in all his studies. He began his University work at Breslau (Figure 6.2) but could not complete it there because Jews were not admitted to the doctor's degree at this university. So, in 1846, Cohn went to the more liberal University of Berlin. Here he studied under Christian G. Ehrenberg (1795–1876), one of the founders of bacteriology and protozoology. Ehrenberg had published his most famous work, *Die Infusionsthierchen als vollkommene Organismen* in 1838. It was Ehrenberg who turned Cohn's attention to the fascinating world under the microscope. In 1847 Cohn returned to Breslau and spent the rest of his life there. He found the road of a Jew not smooth, and when he finally became a professor, he had the title twenty years before the university gave him the laboratory facilities that should have accompanied the title! Finally, the university agreed to the establishment of an Institute of Plant Physiology with Cohn as

Figure 6.2   *The main building of the University of Breslau (Wroclaw) erected between 1728 and 1732. Photographed in 1987.*

the Director, and the Institute was opened on 20 November 1866 in some rooms on the second floor of an old student dormitory. It was the first institute of plant physiology in the world and became a model for similar laboratories established elsewhere.

Ferdinand Cohn was one of the first botanists in Europe to teach from living plants rather than from dried and pressed specimens. The building where his institute was finally established was located opposite the university building itself, and housed a mineralogical museum on the first floor. Cohn's second floor space was hardly attractive—only a high-ceilinged corridor and several poorly lighted rooms—but it was

Cohn's own space and big enough for an enthusiastic researcher and his students (Figure 6.3).

In his early years, Cohn studied algae, but by the 1860s he had turned to a study of bacteria. Finding no microscope available to him from the university, he induced his father to buy him one. Cohn first had to show that bacteria were plants, rather than, as many had insisted, animals (the great German biologist E.H. Haeckel, had called them *Protists*, living organisms that were neither plants nor animals and hence the forerunners of both). Cohn also had to clarify the relationship of bacteria to fungi. He rightly insisted that although bacteria resembled fungi, this

Figure 6.3   *Ferdinand Cohn's Institute of Plant Physiology, on the second floor of an old student dormitory, where Robert Koch made his famous demonstrations of the life cycle of anthrax. Cohn himself is sitting in the background.*

was only in overall living processes and not in kinship. To Cohn, the bacteria (sometimes called the *fission fungi*) were related only to the blue-green algae, which were no more algae than fungi. It is interesting that during the cholera epidemic of 1853, Cohn hypothesized from his studies on suspected drinking water wells that bacteria were the cause of the disease.

From careful observation of bacteria, Cohn concluded that different species of bacteria existed and that a morphological classification was possible. He recognized, however, that organisms that appeared similar in form under the microscope could differ from each other in their physiological characteristics and in their products. One of Cohn's students, Joseph Schroeter, was the first to obtain isolated colonies of bacteria on solid substrates. Schroeter noted that colonies that developed on potato slices were often of different colors, although they might contain morphologically similar organisms, an exceedingly important observation that led to Koch's pure culture plate technique (see Chapter 11). Cohn recognized the taxonomic significance of Schroeter's observations and his concept of bacterial species and bacterial classification was strongly influenced by it.

Cohn established his own scientific journal, *Beiträge zur Biologie der Pflanzen*, and published in this journal an extensive series of papers with the overall title *Untersuchungen über Bacterien* beginning in 1872. Koch's paper on anthrax became one of that series. Cohn also published a prescient and penetrating book with the title *Über Bacterien, die kleinsten lebenden Wesen* (*Bacteria, The Smallest Living Organisms*).[4] This book, which was undoubtedly available to Koch, clearly enunciated the nature and significance of bacteria. A few quotations from this book are given here to indicate Cohn's thinking, and the nature of bacteriological thought when Koch began his work in the 1870s:

"Bacteria form the boundary line of life; beyond them life does not exist, so far as we can tell with our most powerful microscopes."

"If we could view a person under a microscope suitable for examining bacteria, the person would appear as large as Mont Blanc. But even under this extreme magnification, the smallest bacteria do not appear larger than the periods and commas on a printed page. These smallest bacteria may be compared with a human as a grain of sand is to Mont Blanc."

"We are able to discern nothing about the internal structures of bacteria, and even their existence would for the most part be hidden, if they did not live in such large masses."

"Bacteria do not generate the material that forms their bodies *de novo*, but take it from the environment as food. Therefore, no more bacteria can be formed than the food available."

"Bacteria are among the most widespread of organisms. They are omnipresent, in air or water, attached to surfaces, but they only develop into masses when decomposition, corruption, fermentation, or putrefaction can take place."

"In recent times, our knowledge of the effects which bacteria can have over the life and death of humans has been revealed. . . . All epidemics, cholera, pestilence, typhus, diphtheria, variola, scarlet fever, hospital gangrene, epizootic, and the like, have certain features in common. These diseases do not arise *de novo*, but are introduced from another place where they have been prevalent, by means of a diseased person or through material which has been in contact with such: they spread only through contagion."

Cohn discussed Davaine's work on anthrax and other work that suggested that bacteria might cause contagious diseases. "A drop of blood filled with anthrax bacteria will, when introduced into a healthy animal, produce death in 24 to 36 hours, but if inoculation is with blood that contains none of these bacteria, it is without effect. . . . Chauveau and Klebs used filters to show that in pyaemia, septicaemia, and variola, the contagion is not present in the fluid portion of the pus or lymph, but in the microscopical spherical bacteria which are present in it. As these bacteria are strained from the contagious material by the filter, the clear fluid which passes through the filter loses all of its contagious powers, while the particles which remain on the filter are still infectious. All these facts make it in the highest degree probable that many of the bacteria we have already identified are causes of infection. . . . The skill of the physician should therefore be addressed to the question: In what way can the transportation of microscopical organisms be hindered, and by what means can the multiplication of these microscopical organisms be inhibited?"

Cohn wrote, many years later, that when he received Koch's letter he had been very dubious that a completely unknown doctor, living in a Polish section of Germany, would have had anything significant to contribute.[5] Over the years, Cohn had dealt with a number of dilettants who had claimed to have made significant discoveries about bacteria. However, Cohn must have been touched by the respect and innocence shown by Koch's letter and he welcomed him willingly. And within hours after their first meeting, Cohn knew that he was in the presence of a master!

Koch's visit to Cohn

What modesty and humbleness are reflected in the letter which Koch wrote to Cohn:

> Wollstein (Province of Posen)
> 22 April 1876
>
> Honored Professor!
>
> I have found your work on bacteria, published in the *Beiträge zur Biologie der Pflanzen*, very exciting. I have been working for some time on the contagion of anthrax. After many futile attempts I have finally succeeded in discovering the complete life cycle of Bacillus anthracis. I am certain, now, as a result of a large number of experiments, that my conclusions are correct. However, before I publish my work, I would like to request, honored professor, that you, as the best expert on bacteria, examine my results and give me your judgement on their validity. Unfortunately, I am not able to send you preparations which would show the various developmental stages, as I have not succeeded in conserving the bacteria in appropriate fluids. Therefore, I earnestly request that you permit me to visit you in your Institute of Plant Physiology for several days, so that I might show you the essential experiments. If this request is agreeable to you, perhaps you might inform me of a suitable time that I could come to Breslau.
>
> Very sincerely yours,
> Dr. Koch, Kreisphysikus[6]

Cohn immediately agreed to the request and suggested that Koch present himself on the following Sunday. We can well imagine Koch's excitement. But how frantic he must have been, assembling the necessary things for the demonstration. He took with him all his apparatus, containers, reagents, living rabbits, living mice, even frogs. To reach Breslau, Koch took the Post carriage to the train station at Fraustadt. Leaving Wollstein at 1 o'clock in the night, he arrived in Breslau at 10 o'clock the next morning.

Exactly at twelve noon, Koch presented himself at Cohn's house, and then took all his equipment to the Institute. He immediately set himself to arranging the demonstration. We know quite a bit about this historic meeting because Koch kept a diary, and Cohn had a log book at his Institute where all the events and experiments of his Institute were recorded.[7]

Koch's demonstrations

At noon at the Institute of Plant Physiology, Koch took fresh blood from the spleen of a mouse that had just died of anthrax and set up cultures, using his moist chamber with aqueous humor from the veal eye (see Chapter 5). Eduard Eidam, Cohn's assistant, was present as well as Cohn. At 2 o'clock, with the cultures set up, Koch went to lunch by himself, returning at 4 o'clock. Cohn and Eidam were shown how the bacilli grew into long filaments in the slide cultures. At this time, Koch also implanted some of the diseased spleen under the skin of a frog, to demonstrate how the bacilli could develop and penetrate epithelial cells. After all of these experiments were set up, Koch looked around Cohn's Institute, then went to Kissling, a Breslau beer garden, with a friend from the city.

The results of these initial experiments, seen the next day, excited Cohn exceedingly. In the slide cultures, the bacteria from the diseased spleen had grown into long chains of filaments and had formed spores. Since he himself had recently discovered spores in another bacillus, *Bacillus subtilis* (see later), Cohn was especially entranced. "My experiments were well received," wrote Koch in his diary. On this second day, Koch also visited the laboratory supply establishment of J.H. Buchler to look at photomicrographic apparatus, and to the Forchner firm to see equipment for handling animals. He also had time for a walk through the city, where he made some personal purchases. The evening he spent by himself, visiting several local bars, spending more money than he had intended.

By the third day, the importance of Koch's research was known throughout the university. Realizing that Koch's work was of great medical importance, Cohn sent to the Institute of Pathology for someone to come and see Koch's cultures and observations. Because the assistant, pathologist Carl Weigert (1845–1910), was just about to begin an autopsy, the director of the Institute, Julius Cohnheim (1839–1884), went over himself. It was a fateful and dramatic moment, as Cohnheim was the one medical researcher who could not only appreciate the beauty of Koch's work, but could publicize Koch himself.

Julius Cohnheim

Julius Cohnheim, director of the Institute of Pathology at the University of Breslau,[8] was Jewish like Cohn. Cohnheim had been a student of

Rudolf Virchow and was best known for his work on inflammation. When Cohnheim was a student in Berlin, Klebs (see Chapter 4) was one of Virchow's assistants, and Cohnheim was motivated to study pathology from his contacts with Klebs. In 1864 Cohnheim himself became one of Virchow's assistants, carrying out extensive pathological and anatomical investigations, as well as doing teaching and clinical service. This was a time when studies on cellular anatomy were at the forefront and Cohnheim advanced a well-received cellular theory of inflammation. After several other posts and service in the German-Danish war (where his experiences on the battlefield moved him to embrace Christianity), Cohnheim moved to Breslau, where an Institute of Pathology was eventually established for him. Among many studies Cohnheim made at Breslau was one on tuberculosis. Cohnheim discovered the value of the anterior chamber of the eye for the study of infection. In this habitat, the development of suppuration and fluid accumulation that complicated inoculation studies in other parts of the body were absent. Cohnheim had introduced tubercular material in the anterior chamber of the rabbit eye and observed directly the development of tubercles. This work had considerable influence on Koch's later research on tuberculosis, as well as on his selection of the aqueous humor of the eye for his anthrax cultures. Cohnheim became one of Koch's most impassioned champions, and played a major role in advancing Koch's career. Unfortunately, Cohnheim became seriously afflicted with kidney disease, and died in 1884 at the relatively young age of 45.

Koch's continued demonstrations

On the third day, Koch arrived at the Institute of Plant Physiology early and made drawings of his microscopic observations, and he also set up new preparations. The Institute had an incubator which he was able to use. Then the great Cohnheim arrived. Koch showed Cohnheim his cultures and explained his techniques. Together, they took photomicrographs using Cohn's equipment. Cohnheim was tremendously impressed, not only with Koch's experiments, but with Koch himself. When he returned to his Institute, he spoke excitedly to his various assistants:

> Now leave everything as it is, and go to Koch. This man has made a magnificent discovery, which, for simplicity and the precision of the methods employed, is all the more deserving of admiration, as Koch has been shut off completely from all scientific associations. He has done everything

himself and with absolute completeness. There is nothing more to be done. I regard this as the greatest discovery in the field of pathology, and believe that Koch will again surprise us and put us all to shame by further discoveries.[9]

Prophetic words!

Others from Cohnheim's institute, including Weigert, went to see Koch's demonstrations later, and in the afternoon, Koch was invited to visit the Institute of Pathology, where he could see the rooms and laboratories and admire the equipment. William Henry Welch (1850–1934), later to become one of the most distinguished pathologists and bacteriologists in the United States, was a visitor in Cohnheim's laboratory at the time of Koch's visit. Many years later, when Koch was receiving virtually a hero's welcome in the United States, Welch recalled the incident (see Chapter 21). In the evening Koch was at Cohn's house for dinner, together with Weigert and Eidam. By now, Koch had enchanted everyone with his enthusiastic personality and his intimate knowledge of bacteria.

The high point of Koch's visit was the discovery that he and Cohn had independently observed bacterial endospores, Koch with his relatively modest microscope, Cohn with his especially good one. Their observations were so similar that their drawings could not be told apart. One's pictures could be used to illustrate the other's results. Indeed, this is what was actually done when Koch's paper was published in Cohn's journal (see later).

His triumph established and his demonstrations completed, Koch spent the next day relaxing. After finishing his drawings, he packed up his instruments and materials. Then he went shopping in the city and ate lunch at the Hotel du Rome with Eidam. After saying goodbye to Cohn, he visited the Zoological Garden, walking out to the Zoo along the Oder River. He returned to Breslau from the Zoo by river steamer and ate dinner at the Schweidnitzer Keller (a popular restaurant in the basement of the City Hall). The next morning he was up at 5 o'clock and left Breslau on the 6:50 train for Fraustadt. Emmy Koch reported that Robert returned home "radiant". Now full of self-confidence, he turned with renewed vigor to his research.

Cohn and Cohnheim continued to express amazement and wonder at the "iron-clad completeness of Koch's revolutionary methods", as well as the "elegance and definitiveness of his experiments". Can one imagine their excitement. Here was Koch, a complete stranger, with no academic credentials, stunning them with the self-confident way he had put together his research program. Especially Cohnheim, an experimenter of the first rank, was impressed.

Koch's success was not only through the elegance of his research, but also through the force of his personality. Cohn, an especially good judge of character, was charmed by Koch's sympathetic personality. How quickly Cohn was won over is shown by the fact that within two days of Koch's arrival as a stranger, Cohn had invited him for an evening at his house. This gave Koch the chance to see, for the first time, the elegant salon atmosphere of a successful university professor (Figure 6.4). Koch found in Cohn not only a fatherly friend but a valued advisor of immense integrity. Indeed, in Koch's paper on anthrax, published in Cohn's journal, the drawing showing the transition from spores back into bacilli was actually prepared by Cohn, "who was kind enough to prepare this drawing himself with the use of his [superior] microscope" (see later).

> I . . . lay great importance on the fact that, at my request, Professor F. Cohn, to whom I owe special thanks for his trouble, tested and in every respect affirmed my statements regarding the life cycle of Bacillus anthracis . . .[10]

Figure 6.4   *Ferdinand Cohn's study.*

Koch remained in intimate contact with his new-found friends in Breslau. He and Cohn carried on an extensive correspondence about Koch's work and about the anthrax paper (see below), and Cohn's assistant Eidam was sent to Wollstein so that Koch could teach Eidam his methods. Indeed, when Eidam returned to Breslau he began a private course in bacteriology for physicians, the first medical bacteriology course ever taught.

The paper on the etiology of anthrax

As soon as Cohn saw the results of Koch's experiments, he suggested enthusiastically that Koch publish his paper in Cohn's journal, *Beiträge zur Biologie der Pflanzen*. Koch agreed willingly. However, before the paper could be completed, an important control had to be run. As we have seen, Cohn had also discovered endospores, having just found them being formed by a culture which he had isolated from hay infusion.[11] Cohn showed that his spore-forming organism, which he named *Bacillus subtilis*, was resistant to boiling when spores were present, an exceedingly important discovery for the development of reproducible sterilization techniques. Because Cohn's *Bacillus subtilis* formed spores in the manner of Koch's *Bacillus anthracis*, it was important to show that the disease anthrax was linked to Koch's specific organism. Therefore, Koch took back to Wollstein with him a culture of Cohn's *Bacillus subtilis* and immediately set to work to test it for pathogenicity. Cohn's culture did not, of course, cause the symptoms of anthrax in experimental animals, and with this important control out of the way, Koch set himself the task of writing up his work for publication in Cohn's journal.

The famous anthrax paper[12] was finished in three weeks and sent off to Cohn. Cohn was enthusiastic about the paper, but suggested an addition to the title. He wanted to add the words *Untersuchungen über Bacterien* to Koch's title, thus making Koch's paper the fifth in a series of papers on bacteria which Cohn was publishing in his journal. This was quite an honor for Koch, as all the previous papers in the series had been written by Cohn. Paper number 4 was Cohn's own paper on endospores, which appeared in the same issue with Koch's. Indeed, the same plate was used to illustrate both Cohn's and Koch's work (Figure 6.5).

In Cohn's paper he mentions Koch's work with admiration:

Among the bacteria which have been described, the one which is present in the blood of animals and humans with anthrax has special significance, since it is without a doubt significant for the pathology of the disease. In 1875 I noted that since the rod-shaped bacteria as a rule reproduced by resting spores, it would also be expected that the rods of anthrax would also form spores and that these would be the germs of the infection. To my great pleasure, I received a letter from Dr. Koch of Wollstein on 22 April. He had been occupied with studies on the anthrax contagion for a long time and had been finally able to discover the complete life history of *Bacillus anthracis*. He demonstrated this to me at my Plant Physiology Institute . . . I was able to convince myself of the complete correctness of his discoveries . . . I will only remark here that the life history of the anthrax bacillus agrees completely with that of the bacillus of hay infusions. Indeed, the anthrax bacillus does not have a motile stage, but otherwise the similarity with the hay bacillus is so perfect that the drawings of Koch can service without change for the clarification of my observations . . .

Although the paper was written quickly, Koch had to wait patiently for months for it to be published. Among other things, publication was delayed because on 8 August Cohn left for a lengthy trip to England and Scotland. The trip, taken in the service of the Minister of Culture of the German government, had as its purpose a visit to an international exposition of scientific equipment in London. Accompanied by his wife, Cohn took this opportunity to visit many important English scholars. As a world-famous figure, Cohn was welcomed into many noteworthy establishments, including even that of the great Charles Darwin (1809–1882). Cohn also visited the famous physicist John Tyndall (1820–1893), and this visit was of great importance not only for bacteriology but for Koch. Tyndall had just read to the Royal Society his important paper, published the following year[13], on the presence in air of bacteria which

Figure 6.5   *The famous set of drawings of bacterial endospores, illustrating the papers of Cohn and Koch. The original was a color lithograph. Figures 1 through 7 illustrate Koch's paper, the rest refer to Cohn's paper published in the same issue. The legend for Koch's paper follows: Fig. 1. Anthrax bacilli from the blood of a guinea pig. Fig. 2. Anthrax bacilli from the spleen of a mouse, after three hours culture in a drop of aqueous humor. Fig. 3. As Fig. 2, but after 10 hours culture. Fig. 4. As Fig. 2, but after 24 hours. Note the chains of spores. Fig. 5. Spore germination. Fig. 5b was drawn by Cohn to illustrate Koch's paper. Fig. 6. Arrangement for a slide culture. The cover slip has been ringed with olive oil to make it air tight. A warm stage was used to maintain the slide culture near body temperature. Aqueous humor was used for the culture. Even with the naked eye one could see turbidity due to the growth of the long filamentous masses. Fig. 7. Growth of the bacilli in the epithelium of the frog. The bacteria are inside the epithelial cells.*

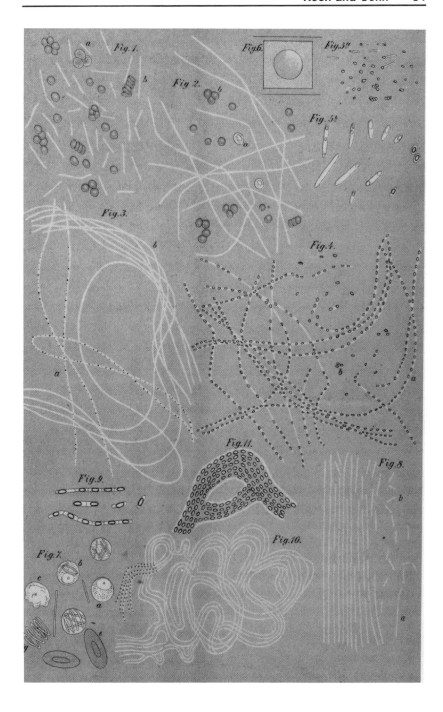

resisted sterilization by boiling. Tyndall's work provided additional evidence for the existence of heat-resistant structures and was obviously of major importance not only for Cohn's studies on hay infusion but for Koch's work on anthrax. Tyndall learned of Koch's work from Cohn on this trip and later became one of Koch's firmest supporters, even going so far as to arrange for publicity and translation into English and publication of Koch's papers.

When Cohn returned from England, he immediately informed Koch about his visit to Tyndall. Koch was by now in the middle of his photomicroscopic investigations (Chapter 7) but wrote enthusiastically:

> I am extremely pleased to learn that my work has received Tyndall's approval.

With Cohn back from England, the next issue of *Beiträge zur Biologie der Pflanzen*, with Koch's paper, could finally go to press. Koch wrote to Cohn:

> Would you think it appropriate for me to send reprints to Pettenkofer, Bollinger, and other editors of medical journals? With this in mind, I would like to request 25 reprints.[14]

Finally, the journal appeared in early October and Koch received his reprints:

> Thank you very much for your letter. I hasten to inform you that I have received the 25 reprints of my paper. In addition to friends, I have sent copies to Pettenkofer, Reklam, and Virchow, as well as to the editors of the German Quarterly for Public Hygiene, the German Medical Journal, and the German Medical Weekly.

Koch's letter offers a clue to the future direction of his work. Although his research obviously had botanical interest, the main importance was medical, and it would be in the medical arena that Koch would need to obtain approval. He might at the moment publish his work in a botanical journal, since this was readily open to him, but Koch was, after all, interested primarily in medical problems. We will see in the next two chapters how rapidly Koch moved into the forefront of medical research.

But now Koch had to contend with the first reactions to his paper by other workers. Cohn had sent a copy of the *Beiträge* to the great Viennese surgeon Theodor Billroth (1829–1894), who responded quickly with a copy of a paper that had been published by his assistant A. Frisch. Billroth concluded that Koch's work confirmed that of Firsch. Frisch had inoculated blood from an anthrax-infected animal under the rabbit cornea (following Cohnheim's technique) and had described the development of "fungus figures". In these same preparations he had observed evidence of structures that Billroth interpreted as spores such as those described by Koch. Because Billroth did not know Koch and lacked his address, he sent a copy of Frisch's paper to Cohn with a request that it be sent on to Koch.

> There is no better indication of the correctness of the observations than the fact that two researchers, starting from quite separate viewpoints, and using different methods, have reached the same result.[15]

However, Koch saw immediately that Frisch's data did not provide evidence for spores. Koch was bitterly disappointed. He wrote to Cohn:

> The only new thing in Frisch's paper is that the anthrax bacilli grow in the living cornea without forming the characteristic filaments. All Frisch's drawings show chains of rods, but no filaments. I do not believe that Frisch has really seen spores . . . Almost certainly, inoculation of animals with these "spores" will not result in anthrax. . . . Frisch's results contradict the results of my extensive studies so greatly that I must assume a confusion with other kinds of bacteria.[16]

But by now, Koch had moved on to his important work on photomicroscopy of bacteria, discussed in the next chapter. Through the efforts of Cohn and Cohnheim, the importance of Koch's paper was gradually appreciated throughout the medical community, although it took several years and several more papers before Koch found complete acceptance of his ideas. His later work at Wollstein, leading up to his important move to Berlin, are discussed in the next two chapters.

7

Koch's Role in the Microscope Revolution

As long as the makers of microscopes do not offer us equipment of higher powers . . . we will find ourselves, when studying bacteria, like a traveler who wanders into an unknown country at the hour of twilight, at the moment when the light of day no longer suffices to enable him clearly to distinguish objects, and when he is conscious that, notwithstanding all precautions, he is liable to lose his way.

—FERDINAND COHN[1]

One of Robert Koch's main contributions was the successful adaptation of the light microscope to the study of bacteria, especially those found in diseased tissues. He was the first to use oil immersion lenses and the Abbe condenser, and he was the first to publish photomicrographs of bacteria. His research on the staining of bacteria for microscopy provided the foundation for this important topic. These remarkable accomplishments were made with equipment and supplies that Koch had to purchase with his own money.

Koch's initial work on photomicroscopy

After his anthrax work, Koch worked hard to obtain better images of bacteria, realizing that his microscope was the limiting factor (Figure 7.1). Much of his work was motivated by a desire to photograph bacteria through the microscope, as he realized that hand drawings were unsatisfactory for communicating the results of bacteriological investigations. His work on photomicroscopy not only forced him to improve

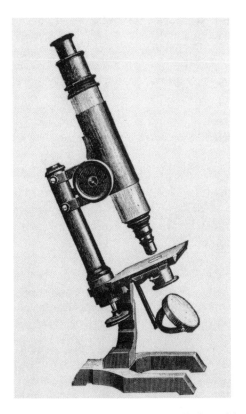

Figure 7.1   *A Seibert microscope, the type Robert Koch used in his first work.*

his microscopy but also to perfect better methods of preparing specimens for microscopy. The slide techniques that have served bacteriology for over 100 years stem directly from Koch's work of 1876–1877. As we noted in Chapter 2, Koch's Uncle Eduard had caused him to become interested in photography as a child. Now, he would use his photographic skills on something really important.

In attempting to obtain better images, Koch had extensive correspondence with Seibert and Krafft of Wetzlar, one of the leading microscope builders of the day.

> July 14, 1876
> I have encountered great difficulty executing proper drawings of bacteria and hope to use photomicroscopy to get around these problems. For years my work on photomicroscopy has been based on the book by Reichardt and Stürenburg.[2] I have been informed by the Institute of Plant

Physiology in Breslau that your optical firm supplies the best apparatus for photomicroscopy. I have obtained your catalog and I do not find among your many models an apparatus that is suitable for my work. I need an apparatus that provides the highest magnification, at least 1200 or even more, but I do not require a large picture size, no larger than 10 cm diameter. . . . I would be exceedingly grateful if you would inform me as to whether it would be possible to obtain good photographs of tiny transparent objects such as bacteria, and if you could advise me which photomicroscopic apparatus would be the best for my purposes.[3]

Koch received an answer and quickly (24 July 1876) ordered the recommended apparatus. He waited impatiently to receive it, firing off a series of letters to Seibert & Krafft inquiring as to the delivery date. Finally, on 2 October he wrote an ultimatum, and a few days later he received his equipment.

10 October 1876

I hasten to write that the photomicrographic apparatus I ordered has arrived in good condition. It does not exactly fit my microscope but I hope that I will be able to adapt it by making several minor alterations. . . . It would please me greatly if you would give me some advice about how the illumination system should be used. How is it used with sunlight? How is the fourth set screw on the shutter used? Is it perhaps necessary to use a piece of ground glass or something similar? In my order I also requested a stage micrometer. Would you please send this to me (1 mm divided into 100 equal parts) as soon as possible? I will then pay you immediately for the micrometer and the apparatus.[4]

Koch's work on photomicroscopy ultimately led to his second paper, also published in Cohn's journal, which contained the first photomicrographs ever published of bacteria (see below). As he continually improved his photographic technique, Koch came to realize that the photographic plate was often better for examining the bacteria in a preparation than direct observation through the microscope. This was because (he said) the light-sensitive plate was not dazzled by bright light, as was the eye, so that small differences in intensity could be seen better on the negative plate than through the microscope. "Often I have easily found fine objects on the negative that I could see only with difficulty through the microscope."[5] At this time he wrote to Cohn:

During the summer I have worked first with anthrax bacilli and later with other Schizophytes, using special methods of preparation to study these organisms precisely, with the goal of distinguishing various species and perhaps accurately telling them apart. My results have exceeded my fondest expectation. I now have a rather large collection of slides of a variety of bacterial forms: Spirillum, several species of spirochetes, bacilli, micrococci, and bacteria. Since it hasn't been possible to show by hand drawings the characteristic arrangement and size relationships, I have decided to prepare photographs. At first I ran into many problems but I believe I have solved the worst of them and I hope to send you in several weeks some photographs of Schizophytes.[6]

However, Koch underestimated the difficulties of getting good photographs. It took him over a year more before this work was ready for publication.

Initially, Koch used a vertical camera-microscope arrangement, as had been described by Reichardt and Stürenburg[7] (Figure 7.2), but this arrangement only permitted a magnification through the microscope of 300 X. Further magnification had to be made by enlargement of the

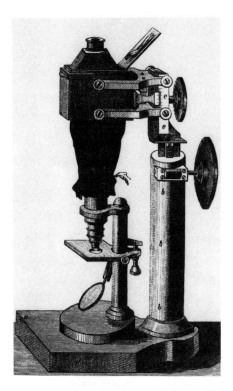

Figure 7.2 *Vertical photomicroscopic apparatus of the type Robert Koch used in his first work, as illustrated by Reichardt and Stürenburg.[2]*

negative, an unsatisfactory procedure. Later, Koch acquired a horizontal microscope-camera setup (Figure 7.3) in which the camera, microscope, and mirror lighting arrangement were carefully aligned on an optical bench. Sunlight was directed onto the microscope mirror by means of a *heliostat*, a device which followed the sun. On the window through which the sunlight was directed, Koch arranged a shutter so that the time of exposure could be controlled by opening and closing this shutter rather than by pulling the dark slide of the photographic plate. In this way, no movement of the microscope occurred when the shutter was opened.

Photography in Koch's time

It is hard for us to appreciate today how primitive the photographic possibilities were in Koch's time (Figure 7.4). Photographic film did not exist and all pictures were taken with emulsion-coated glass plates which the photographer prepared at the time of use. The best images were

Figure 7.3   *Reconstruction of Robert Koch's work room in Wollstein, as shown at a Robert Koch exhibition in Berlin in 1935. The horizontal photomicroscopic apparatus Koch used in his later work is on the left. Note the string-operated shutter that controls the admission of sunlight into the room.*

Figure 7.4   *Arrangement of a photographic darkroom in Robert Koch's time. (a) Door; (b) work table; (c) and (d) arrangement for waste water; (e) window covered with yellow paper to make the room light-safe for photographic work; (f) water storage; (i) shelves holding glass plates being dried; (n) open cassette; (o) tray for the silver bath.*[9]

obtained with wet plates, as dry plates had insufficient sensitivity at the high magnifications needed. We can obtain an idea of how Koch must have worked from the descriptions given in Reichardt and Stürenburg[8] and Gerlach[9].

The glass plates used had to first be carefully cleaned. The photographic emulsion was made from collodion, a solution of cellulose nitrate in alcohol-ether. When spread in a thin film, the solvents evaporated leaving a tough, colorless film. The light-sensitive agent consisted of silver iodide, which had to be formed *in situ* by the photographer. The collodion could be purchased commercially already saturated with iodine.[10] First, the iodized collodion was poured over the glass plate in a thin film, the plate being held vertically to allow the liquid to drain (Figure 7.5a). Then, in a dark room or dark chamber (Koch had a large box built for him by a local carpenter) the glass plate containing the collodion-iodine film was immersed in a silver bath (Figure 7.5b). The silver bath consisted of highly purified silver nitrate dissolved in either distilled water or rainwater. The silver ions reacted with the iodine of the collodion film and tiny crystals of light-sensitive silver iodide were formed. It took several minutes to prepare each plate, and the plate had to be used immediately. For the actual photography, the glass plate

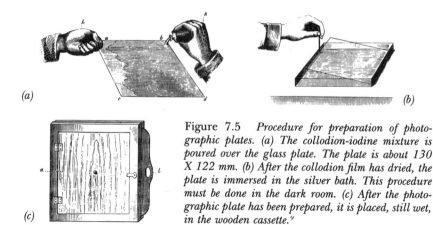

(a)

(b)

(c)

Figure 7.5 *Procedure for preparation of photographic plates. (a) The collodion-iodine mixture is poured over the glass plate. The plate is about 130 X 122 mm. (b) After the collodion film has dried, the plate is immersed in the silver bath. This procedure must be done in the dark room. (c) After the photographic plate has been prepared, it is placed, still wet, in the wooden cassette.*[9]

was placed into a wooden photographic plate holder (Figure 7.5c). Once the cassette was closed one could then remove it from the dark chamber and take it to the photomicroscopic apparatus.

How was the actual photomicrograph made? *Before* preparing the wet collodion plate, the microscope slide had to be placed on the stage and the desired specimen put into focus. Then the photographic plate was prepared in the darkroom. Back at the microscope, the cassette was placed on the photomicroscopic apparatus, the dark slide of the cassette opened, and the exposure made (exposure was generally four to five minutes with ordinary daylight). After the exposure, the dark slide was reinserted and the cassette taken back to the darkroom for developing. In the dark, the plate was removed from the cassette and developed, washed, and fixed. Only then could one examine the plate to see if the exposure was correct and if the specimen had remained in focus.

A typical session of photomicroscopy

The following, taken from Gerlach[11], shows us what Koch had to contend with:

> Before beginning, assess the weather. Only a clear day with a high barometer reading and good sunlight is suitable for taking pictures. . . . Start early in the day, making fresh plates and getting everything ready. It often takes three hours or more to obtain four to six good pictures.
>
> It is best to take the whole microscope apparatus outdoors, rather than

try to shoot through a window, since you will get a lot more shooting time outside.

Make sure the apparatus is firmly mounted so that it does not move when pulling the dark slide in and out. I use a specially made four-legged table of 55 cm height. Clean all the lenses, screw them in completely, and place the illuminating mirror on the sunny side of the microscope. With a dark cloth over your head, look through the ground glass and adjust the light and focus the specimen. . . . Once the image is in focus . . . go inside to prepare the photographic plates. In the darkroom . . . remove a clean glass plate with forceps and pour over its surface the iodized collodion solution, making sure the film spreads evenly and completely. Once the collodion film is ready, close the darkroom door and carefully lower the plate into the silver bath. . . . [After it is ready] allow it to drain and put it in the cassette. Close the cassette and go back outdoors to the photomicrographic apparatus. Remove the black cloth . . . and check to be certain that the proper image is still in focus. . . . Then carefully place the cassette on the apparatus and slowly remove the dark slide from the cassette, being careful not to move anything. After the exposure . . . push the slide back in the cassette, remove the cassette from the microscope, and cover the microscope again with the black cloth. This whole procedure must be done quickly! Run back to the darkroom with the closed cassette, close the darkroom door tightly, take the glass plate out of the cassette, develop the plate, and fix the negative. If the photographic image is not completely sharp, or if there are imperfections in the emulsion . . . it is necessary to repeat the whole process, since nothing is more disheartening in the photographic technique than to try to make prints from unsatisfactory negatives.

We can imagine Robert Koch carrying out the above procedures in between patients! To avoid the problem of running outside, Koch used a clock-operated heliostat which followed the sun and directed the light into his window through a shutter (see Figure 7.3). Then, only Emmy Koch, his little "Wolkenschieber", had to stay outside and warn him when a cloud was about to block the sunlight.

Koch's efforts to perfect photomicroscopy of bacteria

Koch spent the latter half of 1876 and the first part of 1877 attempting to perfect his technique of photomicroscopy. At this time, he viewed photomicroscopy not only as a tool for communication, but as a procedure to aid in the classification of bacteria. He wrote to Ferdinand Cohn on 15 November 1876:

I am working hard to develop a technique that will make it possible to distinguish the various species of Schizophytes, even the smallest and least characteristic. . . . [By making photomicrographs, I can reveal the bacteria] true to nature and free of subjective misinterpretation. . . . Frisch's paper [see Chapter 6] has convinced me even more so of the necessity of improved methods.[12]

However, things did not go well. On 4 December 1876 he wrote:

In order to obtain good negatives at the needed magnifications, one must have very expensive apparatus and bright sunlight. The first is beyond my means and the latter is also unfortunately missing this time of year.[13]

During this period, Koch made contact with several other individuals who were having success with photomicroscopy, including an industrialist named Janisch and Gustav T. Fritsch, a Professor of Physiology at the University of Berlin. They had used a horizontal photomicrographic apparatus, also made by Seibert and Krafft, rather than the vertical one that Koch had obtained the previous summer. Koch then ordered one of these horizontal cameras for himself:

Please don't make me wait as long as I had to last summer before you send me the instrument.[14]

When he finally received the instrument, he worked it over extensively, modifying it for his own purposes (Figure 7.3). Finally, he obtained some negatives which he was satisfied with and sent them off to Cohn. Cohn was extremely excited with the photomicrographs and asked Koch to prepare a paper for his journal. The preparation of this paper occupied Koch throughout the winter, spring, and summer of 1877. The paper not only contained the first photomicrographs of bacteria ever published, but also described in detail all of Koch's procedures, including slide preparation, staining, and preservation of specimens.

The 1877 paper

In his 1877 paper, Koch laid out clearly the precise procedures that he followed when preparing, staining, observing, and photographing bacteria. His methods were described in such detail that others could easily follow them. Koch began by describing how a microscope slide is prepared, beginning with a suspension of bacteria in liquid. Koch emphasized the importance of drying the bacteria-containing fluid in a very thin layer on the cover glass, so that the bacteria were fixed in a single plane. This not only stopped motility and Brownian motion, but also stabilized the sample. (The slide technique for examining cultures still used today differs very little from that first described in Koch's paper.) The preparation was allowed to dry in the air, and such dried preparations could be kept for weeks or months. Koch especially emphasized the importance of this "conservation" technique, since it made it possible for one to keep a bacterial sample for later comparative microscopic study. He noted that drying a preparation prevented the development of contaminating bacteria ("fremder Bakterienarten"). He also described in this paper how he took fresh cover glasses with him to the patient's bedside, so that the patient's fluids could be sampled. (He would later use this technique extensively when searching for the causal agent of cholera, see Chapter 15.) Koch commented on the possible objection one might have to studying dried preparations:

> I saw to my astonishment that bacteria do not collapse and become deformed upon drying, as do infusoria, monads, or algae, but retain their shape, becoming fixed firmly to the glass by way of their outer slime layers without changing either their length or width.[15]

Once dried on a cover glass, the preparation could be later rehydrated with water and stained. The best stains were the aniline dyes. The use of aniline dyes for staining fluids and tissues had been discovered by Cohnheim's assistant Carl Weigert[16], who recommended the procedure to Koch. Important improvements in staining were also made by Paul Ehrlich (1854–1915), Weigert's cousin (see Chapters 14 and 19). Koch noted that these aniline dyes stained bacteria specifically and permitted distinction of bacteria from nonliving precipitates, fat droplets, or other tiny bodies. Koch tried a number of aniline dyes, including methyl violet, fuchsin, safranin, eosin, and methyl green. In some cases, fuchsin worked best but in most cases methyl violet was preferable. For photographic work, where color sensitivity of the emulsion was a consideration, aniline brown was used. Koch described his staining procedures in detail, emphasizing the importance of choosing a proper concentration of dye and the necessity of rinsing well. Once the bacteria were stained, the preparation could be conserved with Canada Balsam or another suitable suspending fluid. Koch's detailed experiments on the staining of bacteria provided a solid foundation for his most important work, the discovery of the tubercle bacillus (see Chapter 14).

Throughout 1877, Koch and Cohn had extensive correspondence, as Koch tried to get his photographs ready for the paper, and Cohn tried to arrange for good positive prints. Koch also travelled occasionally to Breslau, to show his photographs to Cohn and to keep up on the latest news of the bacteriological world. We have a fascinating account of this stage in Koch's life from Carl J. Salomonsen's diary. Salomonsen (1847–1924), a Dane, spent the months from April to August 1877 in Cohnheim's institute and met Koch.[17]

> I must also mention another guest whose visit to Breslau was considered important by both Ferdinand Cohn and Cohnheim: Robert Koch of Wollstein. Koch had achieved world fame as a result of his paper last year that described his research on anthrax. This time he brought with him a series of photomicrographic negatives. According to Koch, all other methods of illustrating bacteria were obsolete, now that photomicroscopy of stained bacterial preparations had been perfected.[18] I was invited to midday dinner with Koch and Cohn; Eidam was also there. We talked about almost nothing but bacteria. Later, we all went to the commercial photographer who was making marvelous prints of Koch's negatives for publication in Cohn's *Beiträge*. Although Koch's visit was brief, he and I remained in friendly contact as "bacteriologists", which was a great aid to me in my later research.[19]

Throughout the spring and summer, Koch and Cohn corresponded frequently about the prints for the paper, arranging the plates, deciding on which pictures to use, commenting on what each photomicrograph showed. At one time, Koch queried Cohn:

> Do you believe, Herr Professor, that my photomicrographs would be suitable for distinguishing the various species/differences of bacteria? The longer I study the Schizophytes, the more convinced I am of my ignorance of these lowest forms of plant life.[20]

Koch also spent a lot of time trying to obtain photomicrographs of what he was calling bacterial flagella, without great success. Finally, he gave up this particular line of work:

> It would be very interesting to follow this line of work further, but it would take me too far from the field of medicine.[21]

Later he wrote:

> I am especially proud of the photomicrographs of Bacillus anthracis, since

it is rare that one finds conditions appropriate to photograph living bacteria by sunlight.[22]

Over this year he worked feverishly on his research, never taking a vacation, seeing patients only when absolutely necessary. It was not until November 1877 that the prints were finished to Koch's satisfaction and the paper could be published. Finally, after one and one-half years of intense effort, the manuscript went to press. Koch had the reprints just before Christmas. He sent reprints to a number of people with botanical and medical interests, including Professor Fritsch in Berlin, to whom he wrote:

> I am well aware of how imperfect my photographic efforts have been but I am absolutely certain that a bad photograph of a living organism is a hundred times better than a misleading or possibly inaccurate drawing.[23]

An examination of a copy of Koch's paper today fills us with admiration. The photomicrographs are, for the most part, outstanding, and would not be out of place even in a modern publication (Figure 7.6). The photos in the journal are not halftones, as are used today, but individually prepared prints. Each photographic print had to be hand-fixed to its place on the plate for each copy of the journal. Examination of these prints with a lens reveals an amazing amount of detail. However, we should note that except for *Bacillus anthracis* (whose large size makes it very favorable for microscopy), all of the samples that had been photographed had been suspended in culture fluid or blood. No tissue slices or pathological preparations were used. Koch was to find out later, when he began to study bacteria that grew in diseased tissue, that the Seibert and Krafft microscope which he used here was not suitable for examination of bacteria in tissue and that photomicroscopy of bacteria in tissues was virtually impossible by the methods available to him.

Ernst Abbe and the development of microscopy

The history of microscopy has been well covered by Bradbury.[24] By Koch's time, achromatic lenses were available, and the principle of immersion had been discovered, but only water-immersion lenses were used, unsuitable for examining such tiny objects as bacteria. The further developments in microscopy, so crucial for the field of bacteriology, are closely linked with the name of Ernst Abbe (Figure 7.7). Abbe (1840–

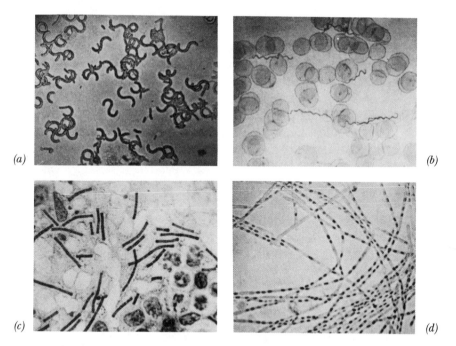

(a)   *(b)*

(c)   *(d)*

Figure 7.6   *A few of Koch's photomicrographs from his 1877 paper. (a)* Spirillum Undula *with "flagella", photographed from a dried unstained preparation. (b)* Spirochaete Obermeieri, *the causal agent of recurrent fever, stained with aniline brown and immersed in glycerol. (c)* Bacillus anthracis, *from infected spleen. A thin layer of tissue was allowed to dry on a cover slip, then the preparation stained with aniline brown and immersed in glycerol. This technique causes the red blood cells to lose their color. (d) Anthrax bacilli which had formed spores after having been cultured in aqueous humor. The preparation was dried on the cover slip, then rehydrated in potassium acetate and photographed without staining.*

1905), a Lecturer in Mathematics, Physics, and Astronomy at the University of Jena, became an optical consultant for the Carl Zeiss Microscope Company. He subsequently (1871) left the University to become a partner and later (1888) the sole proprietor of the firm. Abbe carried out numerous experiments on lens design and acquired a sound grasp of both the theoretical and practical aspects of optics. As a result of his work, Abbe realized the important distinction between *magnification* and *resolution*. He understood that magnification could be readily increased but that this did not necessarily make it possible to *see* anything better. The key requirement was to *resolve* into separate images tiny objects that lay close together.

Abbe showed that in order to resolve closely positioned points, the

Figure 7.7 *Ernst Abbe, the force behind the Carl Zeiss microscope company.*

microscope lens must accept not only the rays coming directly up the axis of the objective, but also at least one of the diffracted beams of light. The more diffracted light which entered the objective, the more faithful would be the representation of the structure. Microscopists of the mid-nineteenth century had discovered that one could increase resolution by using oblique light, light that was directed at an angle up through the specimen. Abbe showed that the reason oblique light increased resolution was because with the axial beam directed into one side of the lens there was a good chance that at least one of the diffracted beams would enter the aperture of the lens. He concluded that by the use of immersion lenses, a much larger aperture could be obtained and therefore a larger amount of diffracted light could be admitted to the lens. Abbe developed the concept of *numerical aperture*, which expressed the light-accepting power of a lens.

Building on his theory, Abbe constructed the first oil-immersion lens. The advantage of oil over water for an immersion lens is that an oil can be used that has the same refractive index as the glass, so that complete homogeneity is obtained, and all of the diffracted light is collected by the lens. Although the concept of *homogeneous immersion* had preceded Abbe, its practical importance had not been realized. Abbe published the first description of the oil-immersion lens in 1879, but his work was known to Koch earlier.[25]

Ever pursuing the better image, in July 1878 Koch traveled to Jena and visited Abbe and the Carl Zeiss Company. On this trip he was accompanied by his Breslau colleague Carl Weigert, and they had the chance to see the Zeiss factory and talk with Ernst Abbe. Koch subsequently obtained one of the first oil-immersion lenses available and used it in his studies on bacteria in wound infections (see earlier in this chapter and in Chapter 8). In Abbe's paper on the oil-immersion lens, Koch's successful use of his lens is mentioned:

> . . . as a proof of excellence of definition which, though indirect, is of special weight, may be mentioned the favourable results which Dr. Koch, of Wollstein, obtained when examining bacteria . . .

These first oil-immersion lenses, supplied by the Carl Zeiss Company, were very successful, and soon they were widely used. They were mainly responsible for the great international success which the Carl Zeiss Company achieved in the microscope field.

However, the oil-immersion lens alone was not enough to ensure superior microscopy of bacteria; one had to provide proper illumination of the microscope field. Abbe's other major contribution was the development of an effective condenser, which came to be known as the Abbe Condenser. In order to provide optimum illumination along the axis of the microscope, Abbe developed a condenser which contained a lens system, so that a full cone of light rays filled the entire aperture of the objective:

> With this illumination, which can only be effected by the aid of a condenser of large aperture, the preparation is simultaneously penetrated in all directions by the incident rays. As a result, the delineation of such parts as stand out in mutual contrast through difference in refractive power (tissue structure, etc.) is almost completely suppressed, and there remain visible only those elements which act as absorbants through staining. . . . Very small and closely clustered elements, as in preparations of bacteria, must certainly . . . become capable of a more thorough resolution than with central illumination of the usual kind.

Koch found that Abbe's condenser was especially valuable when examining stained preparations. Koch made a clear distinction between two kinds of images, which he called the "structure image" and the "color image".[26] Without staining, the structure image was obtained as a result of the diffraction of light, but when staining was used, the color image obtained was much better. Koch recognized that when viewing stained preparations, diffraction of light was undesirable. By the use of the Abbe condenser, it was possible to fully illuminate the field without the attendant problems of diffraction.

It is of interest to note that Abbe and the Carl Zeiss Company provided the important equipment needed for Koch's work, and in turn that it was Koch's success that helped to establish the Carl Zeiss Company as the preeminent microscope builder of the world.

However, even before he obtained the new Zeiss equipment, Koch had returned to work on experiments involving medical problems. Following the lead of Lister's antiseptic surgery, Koch began to study the important problem of wound infections, using animal models. We discuss this important work, so pivotal for Koch's own career, in the next chapter.

8

Studies on Wound Infections: the Later Wollstein Years

*I see from your letter that recently many new things have been reported
which I, in my isolated corner of Germany, know nothing about. Well,
there is nothing for it but I must come to Breslau on a visit and learn at
first hand of all these exciting things. But I am determined not to come
empty handed! I have had a bit of luck in my work, and my general
conclusions seem to me to be fairly important. . . . I have succeeded in
adapting the microscope so that the way is opened to be able to see the
smallest organisms with certainty in animal tissues. You see, honored
professor, that I am promising much, but I am certain of my facts and will
make good on my promises.*

—Robert Koch[1]

Throughout the period after Koch's initial success with anthrax, his self-
confidence increased enormously. All his spare time was now occupied
with research, and he continued to show new and innovative approaches. We discussed his work with photomicroscopy in the previous
chapter. Throughout the period of a year and a half during which he
attempted to obtain useful photomicrographs of bacteria, he had an
extensive correspondence with Ferdinand Cohn about many subjects
(Figure 8.1). That he was isolated in Wollstein became of increasing
concern:

> 15 July 1877
> Many thanks for sending me the bacteriological literature [32 reprints of
> papers that Cohn had been sent by others] and the news articles from
> the Schlesische Presse and the Neue Freie Presse. The information about
> Pasteur's cultivation of the anthrax bacillus is very interesting. If I only
> could study Pasteur's work in the original French.[2]

Cohn tried to find Koch a position in Breslau, to alleviate his difficult

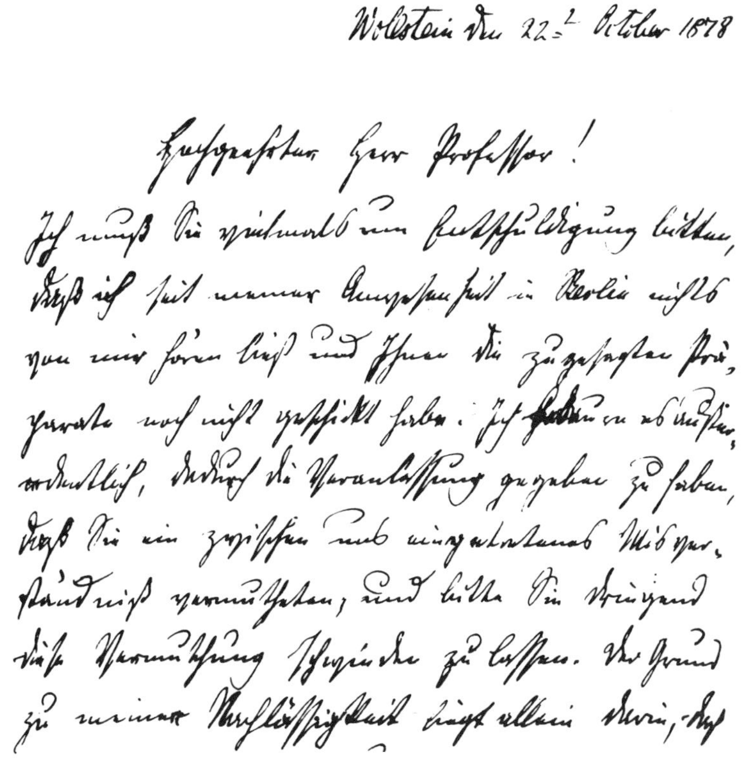

Figure 8.1 *Part of a letter from Koch to Cohn.*

scientific isolation. However, nothing came of it in the fall of 1877 or at several later dates. When Koch finally did move to Breslau, in the summer of 1879, it turned out to be a big mistake (see Chapter 9).

By the fall of 1877, the only paper Koch had published was the one on anthrax, which was eliciting both interest and controversy. An early English proponent of the germ theory of disease, John Burdon Sanderson (1828–1905) (at that time Professor of Physiology at the University of London) wanted to see Koch's experiments for himself, and Cohn arranged for them both to be in Breslau at the same time. Koch packed up all his equipment once again and came to Breslau on 15 October 1877. Among other things, he brought with him an infected mouse which had been injected with blood that had been dried for five years. Koch's itinerary took him by way of Rakwitz and Tarnowitz (where he visited his brother Hugo), and the mouse died on the way. Koch showed Burdon Sanderson the massive development of anthrax bacilli in the spleen of this mouse, and set up cultures in aqueous humor. Later, the cultures sporulated. Koch also used some of the spleen of the dead mouse to inoculate rabbits, which died 24 hours after inoculation. Attending these demonstrations were also Eidam, Cohnheim, Weigert, several other Breslauer physicians, and Albert Neisser (1855–1916), who several years later would first observe the causal agent of gonorrhoea. Also present was the American William Henry Welch, who was on his first German visit. In later years, Welch wrote about his so-called *Breslau summer semester of 1877*: "I treasure with keenest delight the memory of the wonderful days in Breslau."[3]

Indeed, Breslau at this time was *the* center of bacteriological research in Germany and Koch valued his Breslau visits very much. His visits gave him a chance to meet people, to discuss his research, and to gain further confidence. Burdon Sanderson was already strongly committed to the germ theory of disease, so he did not take much convincing, but Koch returned home full of zeal, starting new animal inoculations the very evening of his return. He wrote Cohn an enthusiastic letter:

> As you know, I have no further ambition than to work as hard as I can for science, but the words of praise which have been offered give me some evidence that my efforts, as little as they have been, are worthwhile, and that the direction that I have taken in my research is the correct way.[4]

This expression of personal dedication tells us much about Koch and what was motivating him. Considering Koch's later detractors, and the tremendous criticisms that he suffered (see especially Chapter 18), we can see here, in a nutshell, the essence of Koch's personality. Most scientists are driven to succeed, but rare is the scientist as strongly motivated to excel without regard to fame and fortune as Robert Koch was at this time.

Nägeli's work and the question of bacteriological species

At this time, one of the most important bacteriological questions was whether distinct species of bacteria existed. On one side of this argument was Ferdinand Cohn (and later Robert Koch), who insisted on the specificity of bacterial types. On the other side was the Swiss botanist Carl von Nägeli (1817–1891), who resisted the notion of bacterial species distinctions.[5] In 1877 he published a book[6] which was a strong attack on Cohn's whole approach to bacteria:

> Recently Cohn has constructed a taxonomic system for the bacteria, with a number of distinct genera and species. In Cohn's system, each function of a lower fungus is used to define a separate species. He has thus legitimized an idea that has widespread support, even among medical workers. However, the actual basis by which morphological or other characteristics can be used to separate species is a mystery to me. *For over 10 years I have examined thousands of different fission organisms and (with the exception of the Sarcinae) I have been completely unable to distinguish even two distinct species.* [italics added][7]

Koch was furious when he read Nägeli's book and dashed off a letter immediately to Cohn:

> I have just read Nägeli's book. . . . Seldom have I read a book that has so many errors and so much nonsense in it. The experiments reported are inappropriate for the questions being asked, since he only has mass cultures (*Massenkulturen*).[8]

Why were Cohn's taxonomic ideas so important for medicine? If it could be shown that one bacterium could turn, in an unpredictable manner, into another, then it would be difficult to accept the fact that a *specific* disease was caused by a *specific* bacterium. Koch's work on anthrax was strongly based on the idea that a given disease was caused by a *single* organism. Indeed, in the case of anthrax, it was relatively easy to prove that this was so, since the bacterium was large and morphologically distinct and was not found in hay infusions or other cultures derived from saprophytic sources. But as Koch realized early, the crux of the matter was an animal experiment with a pure culture, always with careful microscopical control. Nägeli and others of his school were casual, even sloppy, in their culture work, and were also inexpert at using the microscope to look at organisms as small as bacteria. Koch, completely on his own, had developed the exactly appropriate techniques.[9]

The etiology of wound infections

By late 1877 Koch had finished his paper on microscopy of bacteria and had returned to specific medical problems. Now he became occupied with an exceedingly important problem, the growth of bacteria in wounds and surgical incisions, commonly called *sepsis*. This work ultimately led to the publication of a small book which was to find favorable acceptance.[10] At this time, terms such as *pyemia, purulent infection, putrid infection, septicemia,* and *traumatic fever* were all used to refer to conditions which, we now know, arose from the growth of bacteria in the body. Among the most important of these conditions was *surgical sepsis*. Joseph Lister, through his development of antiseptic surgery, had already provided strong impetus for considering these conditions to be bacterial infections, but experimental data to this effect were not strong. C.J. Davaine, the French scientist who had been Koch's forerunner in studies on anthrax (Chapter 5) had shown in 1872 that if putrid blood taken from a septicaemic patient was injected into a rabbit, a condition similar to septicemia could be induced. Davaine carried out a series of consecutive injections through 25 rabbits, and observed that each rabbit in the series died following the injection. He calculated that after the 25th passage, only 1 billionth of the original blood was present and concluded (erroneously, see later in this chapter) that the contagion of septicemia acquired an enormous increase in virulence upon animal passage. Davaine wrote of the "ferment of putrefaction" as the agent responsible for septicemia, but it was not clear if he meant "ferment" in Pasteur's sense of a living agent, since Davaine did not carry out microscopical investigations.

Microscopy of wound infections was carried out by the German Edwin Klebs independently of Davaine. Taking advantage of the availability of numerous patients with gunshot wounds at a battlefield hospital during the Franco-German war of 1870, Klebs used the microscope to examine fresh and preserved specimens from over 100 autopsies and found bacteria of different forms in nearly every case.[11] However, medical opinion at that time held that all bacteria were derivatives of a single organism,[12] so that Klebs misinterpreted his microscopic observations and described all the bacteria as one organism to which he gave the name *Microsporon septicum*.

The way was thus open for Robert Koch to apply his careful microscopy and animal experimentation. Building on the animal work of Davaine and the microscopic work of Klebs, Koch carried out extensive investigations at Wollstein and developed the theory that each septic condition was due to a different organism. The importance of this work has been emphasized by Bulloch:[13]

These modern views really started with Robert Koch in his epoch-making work on the *Aetiology of Traumatic Infective Diseases*. This small work of eighty pages, written while Koch was still in medical practice in Wollstein and far from academic influences, was totally unlike anything that preceded it on the subject of septic diseases. His object was to determine whether the infective diseases of wounds are of parasitic origin or not. He had to confine himself exclusively to animal experiments, but he was able to show, in a manner practically conclusive, that a series of diseases, differing clinically, anatomically, and in aetiology, can be produced experimentally by the infection of putrid materials into animals. He admitted that the work of many previous investigators had rendered the parasitic nature of disease probable. Conclusive proof had not been obtained and in Koch's opinion could only be obtained by "finding the parasitic microorganisms in all cases of the disease in question, when we can further demonstrate their presence in such numbers and distribution that all the symptoms of the disease may thus find their explanation, and finally when we have established, for every individual traumatic infective disease, the existence of a microorganism with well defined morphological characters."[14]

Koch's investigations on wound infections

Koch began by following Klebs' lead on microscopy. However, being influenced by Cohn's ideas of bacterial species, and his own experience with microscopy of bacteria, Koch did a much more careful job of studying his materials:

> According to my own experience, the examination of blood for the possible presence of bacteria is exceedingly difficult, unless one uses the special microscopic procedures that I describe below.
>
> I have, on many occasions, examined normal blood and normal tissues using methods that ensure that such organisms are not overlooked, and I have never, in a single instance, found bacteria. *I therefore conclude that bacteria do not occur in the blood or tissues of healthy animals or humans.* [italics in original]
>
> On the other hand, some of the objections to the conclusion that bacteria do cause traumatic infective diseases are well founded. In order to prove that bacteria are the cause of traumatic infective diseases, it would be absolutely necessary to show *that bacteria are present without exception and that their number and distribution are such that the symptoms of the disease are fully explained.*[15] [italics in original]

As noted, careful microscopy was one of the keys to Koch's success. When examining diseased tissues for the presence of the tiny bacteria associated with wound infections, however, the methods which worked so well with anthrax were inappropriate. Koch's colleague at Breslau, Carl Weigert, had discovered that bacteria could be much more readily observed in animal tissues if they were stained with aniline dyes. Koch graciously acknowledged his debt to Weigert for communicating his results to him before publication.[16] We see here another advantage to Koch of his "Breslau connection".

As discussed in Chapter 7, Koch's work marked the entry of the Abbe condenser into bacteriology and medicine. Abbe had first described his condenser in 1873 but it had not found favor in pathology or histology. When Koch began to examine diseased *tissues* for bacteria, he experienced great difficulty with the microscopes that were available to him. He traveled with Weigert to Jena to visit Abbe and the Zeiss factory, where he saw the Abbe condenser:

> Only with the aid of the Abbe condenser have I been able to see bacteria in blood of septicemic animals. This is all the more amazing when it is considered that these bacteria are extremely tiny. . . . Everywhere in the diseased body micrococci were visible . . .[17]

Only after Koch's work did the Abbe condenser become well known and widely used. The impact of this simple development on the field of medical bacteriology is inestimable. The other development which Koch saw in Jena was the oil-immersion lens. These lenses were not available commercially until January 1879, but Koch was able to test one to see if it would be suitable for his purposes during his trip to Jena in July 1878.

> [The oil-immersion lens] completely altered the pictures. In the same slides which had previously shown nothing, the smallest bacteria are now visible with such clarity and definition that they are very easy to see and to distinguish from other colored objects. Even more—what had been up to then only a pious hope—it is extremely easy to see pathogenic bacteria in tissues and to be able to distinguish them taxonomically. Whereas previously all we could see were micrococci and zoogloea [a term commonly used at that time for amorphous clumps of bacteria], now we can see bacteria which can be differentiated by size and shape.[18]

Now that he had the microscopy of diseased tissues perfected, Koch could turn to the use of this technique in the study of wound infections. Following Davaine, but with much better microscopical control, he set up experimental infections in animals. He used mice primarily, injecting them with low doses of diseased material and then making successive animal passages. In one particular experiment, he made 17 successive passages, noting that death always took place in about 50 hours, and always with the same symptoms (which he carefully describes). He used house mice in these experiments; he had tried rabbits and field mice but they were not susceptible. (As we noted earlier, Koch did not have commercially available mice. At first he trapped his own experimental animals in the horse barn or field, and later he raised white mice.)

One of the strengths of Koch's little book is that it was well illustrated with careful drawings which he made of various experiments. Some of his drawings are reproduced here as Figure 8.2. Koch worked hard to obtain good photomicrographs to illustrate his book, but was unsuccessful. He laid out the problem clearly in the Preface to his book, explaining that the difficulty was primarily because of the lack of color sensitivity of the photographic emulsions available to him. Seeing the bacteria depended upon observing color differences between the stained bacteria and the surrounding tissues, but the emulsions available in those days were only sensitive to the blue region of the light spectrum. In order to render the bacteria visible for photography, Koch had to use a light filter made from eosin and collodion. However, because this light filter also reduced considerably the intensity of the light reaching the microscope specimen, very long photographic exposures had to be made, and the unavoidable vibration of the apparatus led to images that were not sharp.

After his book was published, Koch did succeed in obtaining satisfactory photomicrographs of diseased tissues. He sent some of these photomicrographs to Lister, who made them available to W. Watson Cheyne, the English translator of Koch's book. Cheyne notes in his translator's preface that these photographs "show plainly that the drawings are faithful representations of what has been seen".[19] In all, Koch's book included five Plates, each of which contained four or five separate drawings. The drawings were made using a camera lucida and were reproduced as two-color lithographs.

Koch studied a wide variety of traumatic infective diseases, including tissue gangrene in mice, spreading abscess in rabbits, pyemia in rabbits, septicemia in rabbits, and erysipelas in rabbits. In addition, he again studied anthrax in mice, using his new microscopic equipment. The work seems rather simple from our modern vantage, and it is hard to appreciate its significance. Not only did it provide important insights into the bacterial nature of these important conditions (disease con-

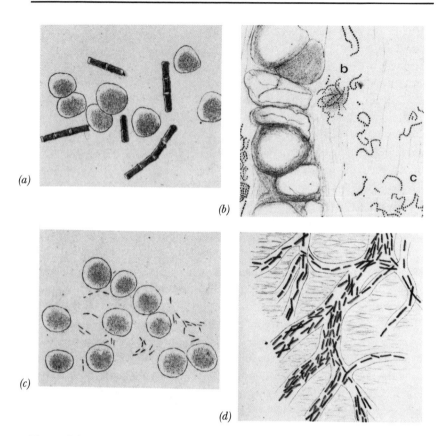

Figure 8.2   *Koch's drawings of bacteria in wound infections. (a) Blood of a mouse infected with anthrax. Red blood cells and anthrax bacilli. (b) Portion of cartilage and adjacent tissue of a mouse ear, showing chain-like masses of micrococci (labeled* b *and* c*). (c) Blood of a septicaemic mouse, dried on a cover slip, stained with methyl violet, and mounted in Canada balsam. Red blood cells and small bacilli are seen. (d) A part of the vascular network of a septicaemic rabbit.*

ditions so much more important then than now), but the microscopical methods and animal experimentation provided the basis for work by Koch and others on the etiology of a wide variety of infectious diseases. The work provided the essential foundation for Koch's later work on tuberculosis (Chapter 14).

However, Koch had not as yet done any serious cultivation of these organisms. He recognized the importance of pure culture studies, although at this time he valued such studies primarily to prove that each pathogenic bacterium was a distinct species:

The greatest stress . . . is justly laid on the so-called pure cultivations, in which only one definite form of bacterium is present. This importance is seen because if one and the same form of bacterium is always obtained in a series of cultivations, then a special significance must be attached to this form. It must be, in a word, a species. Can, then, a series of pure cultivations be carried out without admixture of other bacteria? In truth, only under very limited conditions. Only such bacteria can be cultivated pure, with the methods presently available, that can be easily recognized as pure, either because of their large size, as the Bacillus anthracis, or by the production of characteristic colors, as the pigment bacteria. . . . But the case is quite different when one attempts to cultivate the very small bacteria which are associated with traumatic infective diseases, since these bacteria cannot be seen except by staining. How are we then to discover the occurrence of contamination? It is impossible to do so, and therefore all attempts at pure cultivation are subject to error and will be inconclusive.[20]

Koch had not yet developed his plate technique for pure culture isolation. It was not until he had available all the facilities of his Berlin laboratory, several years later, that he was able to develop reproducible methods for the important and difficult problem of cultivation outside the animal body. But he recognized the possibility of an alternate approach:

But nevertheless a pure cultivation is possible, even in the case of the bacteria which are smallest and most difficult to recognize. Such cultivation is conducted not in laboratory apparatus, but in the animal itself. . . . In fact, there exists no better cultivation apparatus for pathogenic bacteria than the animal body . . .[21]

Another important consideration from this work, so vital for Koch's work on tuberculosis a few years later, related to the possibility that animal passage might "select" bacteria that were more pathogenic than the original. Davaine had interpreted his animal experiments in this light. Koch concluded his book with a discussion of this so-called "law of increasing virulence":

The discovery of this law has, as is well known, been received with great enthusiasm, and it has excited no little interest owing to its intimate bearing on the doctrine of natural selection [*Anpassung und Vererbung* in German].[22]

But Koch pointed out that because the bacteria grew in the tissues, they increased enormously in numbers in the first animal passage.

Therefore, the bacteria used to inoculate a second animal were mostly new bacteria and were not those derived from the original diseased tissue. The same could be said for the third and subsequent passages. And since the animals even in the early passages died as rapidly and from as small doses as those in subsequent passages, there was no evidence of a continued increase in virulence. Koch also recognized, in such animal passages, that one was using the animal as a means of "selecting out" the pathogen from the original mixture, a procedure which would many years later come to be called an "enrichment culture". Koch's writing in his book reveals that he clearly understood the nature of enrichment culture in 1878.

At this point, we might summarize the three major advances that were presented in Koch's 1878 book:

1. Staining of bacteria in diseased tissues with aniline dyes

2. Introduction of the Abbe condenser

3. First use of the Zeiss oil-immersion lens

Ogston's work

But the connection of Koch's work on sepsis to humans was lacking. Certainly Koch would have worked with humans if he had been in a medical center, but isolated in Wollstein, he lacked the necessary clinical material. The connection with humans was quickly made by the Scottish surgeon Alexander Ogston (1844–1929, knighted in 1912)[23]. Following the methods from Koch's 1878 paper exactly ("I am able to affirm the correctness of his assertions. His processes were employed in the following . . ."[24]), Ogston showed in 1880–1883 that two different micrococci were present in massive numbers in pus. Using careful microscopy, amazingly accurate counting methods, and animal inoculation, Ogston showed that there were two kinds of micrococci in pus, one arranged in chains, which had been called *Streptococcus* by Billroth, and one arranged in masses, which Ogston called *Staphylococcus* (from the Greek, a bunch of grapes) and that these organisms grew in the body and were responsible for the pyogenic condition. Ogston's publications exerted an important influence on surgery, pathology, and bacteriology and served to further advance Koch's reputation.

Koch's trip of Summer 1878

Koch pursued his work on septic infections throughout the spring and summer of 1878. Edwin Klebs was also actively engaged in bacteriological studies and he and Koch got into rather vigorous arguments. Klebs' microscope preparations were of very poor quality, which made the debate even more touchy. During this time, Koch was carrying on an extensive correspondence with Cohn. Cohnheim and Weigert had moved from Breslau to Leipzig, where Cohnheim had been appointed Professor of Pathology. Koch felt increasingly isolated and arranged to take a trip and visit the various research centers in Germany that were working on bacteriological problems. He wrote to Cohn:

> 27 June 1878
>
> I see from your letter that recently many new things have been reported which I, in my isolated corner of Germany, know nothing about. Well, there is nothing for it but I must come to Breslau on a visit and learn at first hand of all these exciting things. But I am determined not to come empty handed! I have had a bit of luck in my work, and my general conclusions seem to me to be fairly important. . . . I have succeeded in adapting the microscope so that the way is opened to be able to see the smallest organisms with certainty in animal tissues. You see, honored professor, that I am promising much, but I am certain of my facts and will make good on my promises.[25]

Koch began his trip with a visit to his old friends Cohnheim and Weigert in Leipzig. Cohnheim was very enthusiastic about Koch's new work on the bacteriology of sepsis and strongly urged him to write a short book describing his methods and results. Leipzig at that time was one of the main publishing centers of Germany and Cohnheim had no difficulty in finding a publisher for Koch's book. The details of the book have been discussed earlier in this chapter.

After Leipzig, Koch went on to Breslau, where he conferred with Cohn and demonstrated his new preparations to many others. From Breslau he took the overnight train to Berlin, arriving on Saturday morning and immediately going about his visits. First he visited Gustav Fritsch, a physiologist with whom he had been corresponding about photomicroscopy (see Chapter 7). At this time, Koch was trying without success to take photomicrographs of bacteria in pyogenic infections and diseased tissues, and he had hoped (vainly, as it turned out) that Fritsch could help him.

The same day, Koch went to visit the great Rudolf Virchow, a visit

Figure 8.3   *A view of Koch's primitive laboratory facilities in Wollstein.*

presumably arranged by Cohnheim, who had been Virchow's student. The visit with Virchow was very unsatisfactory. Virchow received Koch cooly and was quite unimpressed with his preparations and discoveries. Concerning Koch's paper on anthrax, Virchow announced that the "whole business seemed quite improbable". When Koch attempted to describe his new microscopic procedures using the oil-immersion lens, Virchow announced that "anything he couldn't see with a dry lens wasn't worth looking at". It has been reported that Koch left Virchow's laboratory "with bitterness in his heart". [26]

This was the second meeting of Koch and Virchow, the first having been in 1875 when Virchow visited Wollstein to examine some archaeological diggings (see Chapter 4). Virchow was to see Koch soon in a different light. Only *four years later*, Koch gave his famous address at the Berlin Society of Physiology on the etiology of tuberculosis and was catapaulted into world fame (see Chapter 14).[27]

Shortly after Koch returned to Wollstein, the Leipzig publisher C.F.W. Vogel brought out his book on wound infections that we have discussed above. The work appeared with a rapidity that could amaze us even in this day of rapid publication.

Cohnheim also convinced Koch to present a lecture on his work at the *51st Versammlung Deutscher Naturforscher und Ärzte* meeting in Kassel 11–17 September 1878. Despite the cost, which he had to bear out of his own pocket, Koch agreed and gave a paper in the Section on Pathological Anatomy and Internal Medicine, primarily in the form of a microscopic demonstration. This demonstration, which was summarized in a brief written report[28] seemed at the time to have little impact on the medical or scientific world.

Returning home from the Kassel trip, after having been away also in early August, Koch found himself swamped with medical duties and was increasingly unable to find time for research work. Obviously, he needed a better position than that of *Kreisphysikus* in Wollstein (Figure 8.3). Soon his situation would break and he would be receiving the professional recognition which he deserved. But in the meantime, he had almost another year to spend in Wollstein!

> We thus come to an end of a chapter, in pathology and bacteriology, which long remained barren in its results but was ultimately clarified by the technical methods pursued with such diligence and with such mastery by Koch and his successors.[29]

9

On to Berlin

By the selection of Robert Koch I have in mind not only his extensive experience in medical practice, but his high skill in experimental pathology and microscopy. . . . It is therefore a happy accident that Herr Koch, one of the most experienced researchers in this area, and a competent, hardworking, and experienced scientist, can be easily attracted to this position.

—HEINRICH STRUCK[1]

It had become increasingly apparent to Koch that it would be impossible to satisfy his scientific aspirations in Wollstein. It was certainly unsatisfactory to try to squeeze research studies in between patients—at times the demands of his medical practice left him exhausted. Also, he needed a better-paying position, as his research expenses were mounting. Fortunately, circumstances eventually made it possible for him to obtain a position that would make the best uses of his talents. By July 1880 Koch was in Berlin, beginning an exciting new position. But first, he made an unsatisfactory move to Breslau.

Koch in Breslau

In January 1879 the medical faculty of the University of Breslau, acting at Cohn's insistence, put forward to the Minister of Culture the suggestion that Koch be appointed an *ausserordentliche* professor of hygiene or public health (this was a professorship but without many of the rights and perquisites of a designated chair). The letter of nomination included the names of distinguished professors of the medical faculty, as well as the Dean and the Rector of the University:

> Under the most difficult circumstance, isolated from other scientists, Koch has made outstanding discoveries. Such a powerful individual would make enormous contributions to the university. At the same time, Herr Koch has great difficulty finding time and energy to carry out his researches in his present position.[2]

However, there was only one problem with this request: there was no available position. The request regarding Koch was coupled with an extensive plan of the medical faculty to expand the programs of the university by the establishment of an Institute of Hygiene, within which Koch's position would be located. However, this proposal was much too ambitious and was premature (the proposed institute was established only many years later).

Koch himself only found out about the proposal two months after it had been made. Declaring himself "happily surprised", Koch indicated he would have been quite pleased with such a plan had it worked out, as he eagerly wanted to move to Breslau. Cohn and his colleagues then proposed that Koch be appointed the *Gerichtliche Stadtphysikus* for the city of Breslau, a position somewhat analogous to his position in Wollstein, but in a much bigger city and with better pay. After much worry about whether this was the right move, Koch agreed to take this position and moved with his family to Breslau in July 1879. It was a sad day for the Wollsteiners, who liked their physician very much and were quite proud of his scientific successes.

In Breslau, Koch rented the 2nd floor of a large apartment building (Gartenstrasse 40a), five rooms with a view of the courtyard and garden, including a good work room for his microscope and experimental animals. Unfortunately, he found it difficult to establish the private medical practice that was essential for him to support his family. Even with his increase in salary as *Stadtphysikus*, he was unable to make ends meet. Emmy and Gertrud also found Breslau strange and difficult and longed to return to Wollstein. Within a few weeks, Koch knew he had made a mistake.

Fortunately, his old position in Wollstein was still open and in October 1879 the family returned. Even their old house was still vacant. The happy people of Wollstein met Koch on his arrival and escorted him home with a torchlight parade.

However, Wollstein's joy was to be short lived. Koch's scientific fame was now so widespread that he would never be able to remain a "lone researcher". In July 1880, Koch moved to Berlin, the head of a newly established laboratory for bacteriological research.

The Imperial Health Office

When the German states were brought together by Otto von Bismarck in 1871 as the German Empire (see Figure 2.1), Berlin became its capital (see Chapter 14 for a brief history of Berlin). Among the administrative departments established was a central office of public health, the *Kaiserliche Gesundheitsamt* (Imperial Health Office). Within a few years, this office became established in its first headquarters, a rented building at Luisenstrasse 19, close to the giant Charité hospital (Figure 9.1; see also Figure 3.1). The first Director appointed to head this office was Dr. Heinrich Struck, whose main claim to fame seems to have been that he was the personal physician of Chancellor Bismarck. The personnel consisted of several professors who were appointed adjunct members of the *Amt*. At first, the office served primarily in an advisory capacity, but soon it was decided that the *Amt* should carry out its own research on problems of public health, and two laboratories were established, one for chemistry, the other for hygiene. Initially, the Laboratory of Hygiene was under the direction of Dr. Gustav Wolffhügel (1854–1899), who had been a *Privatdozent* at the University of Munich. Since the rented quarters were not large enough for a laboratory, a small, rather old, private house at Luisenstrasse 57 was purchased in late 1879. This was to become Robert Koch's first laboratory in Berlin (Figure 9.2).

The direction of the *Amt* was under the guidance of an advisory council consisting of 16 members from the various states of the German Empire, and representing the various disciplines of medicine and public health. Among these council members was Ferdinand Cohn, who was placed on the council to represent "mycology", and Robert Koch, whose work on bacteria was by now quite well known. Koch, although remaining in Wollstein, was appointed to the council in January 1880, presumably at Cohn's suggestion.

By March 1880, Struck was working to obtain for Koch a position as a regular member of the *Amt* staff:

> By the selection of this man, I have in mind not only his extensive experience in medical practice, but his high skill in experimental pathology and in microscopy. The lack of such experience at the *Gesundheitsamt* is

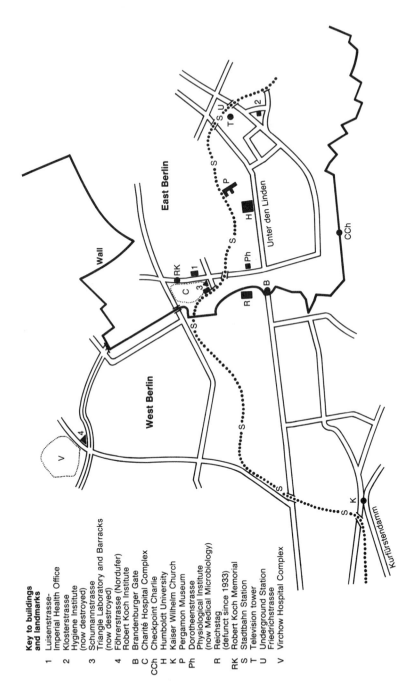

**Key to buildings
and landmarks**

1 Luisenstrasse-
   Imperial Health Office
2 Klosterstrasse
   Hygiene Institute
   (now destroyed)
3 Schumannstrasse
   Triangle Laboratory and Barracks
   (now destroyed)
4 Föhrerstrasse (Nordufer)
   Robert Koch Institute
B Brandenburger Gate
C Charité Hospital Complex
CCh Checkpoint Charlie
H Humboldt University
K Kaiser Wilhelm Church
P Pergamon Museum
Ph Dorotheenstrasse
   Physiological Institute
   (now Medical Microbiology)
R Reichstag
   (defunct since 1933)
RK Robert Koch Memorial
S Stadtbahn Station
T Television tower
U Underground Station
   Friedrichstrasse
V Virchow Hospital Complex

Figure 9.1   *Map of central Berlin, showing some of the key streets and landmarks
and the locations of Koch's various laboratories.*

Figure 9.2   *Laboratories of the Imperial Health Office in Berlin, Luisenstrasse 57. Koch's laboratory was located here until 1885. The building still stands, although the street is called Hermann-Matern-Strasse (East Berlin). It now houses the Central Institute of Library Science of the German Democratic Republic. The photograph was taken in 1987.*

> serious. It is therefore a happy accident that Herr Koch, one of the most experienced researchers in this area, and a competent, hardworking, and dedicated scientist, can be easily attracted to such a position.[3]

Permission to recruit Koch was quickly received and on 9 April 1880 Struck first approached Koch. However, the position Struck had engineered at this time was primarily honorary, and Koch would still have had to practice medicine in Berlin to make ends meet. Koch therefore raised the question of a position with salary, so that he might be able to "feed my family". Struck returned to the Minister and requested a salaried position and in early July, the Minister finally agreed. For the rest of his life, Struck took great pride in telling everyone that it was he, Struck, who had discovered Koch and brought him to Berlin.

Struck telegraphed Koch in Wollstein on 7 July and offered him the position:

> Please reply immediately on your starting date.[4]

Koch needed no time to think *this* offer over. He telegraphed back:

> I will be at your disposal in Berlin on 10 July.[5]

Three days later!

On 9 July, Koch and his family moved to Berlin. They took practically nothing with them, having a quick auction of all their furniture. Even Koch's dark box, some other photographic equipment, and his incubator were sold.

In honor of Koch, the city of Wollstein later placed a plaque on the house at Number 12 *Strasse am weissen Berge* (originally in German, but in Polish since World War II; see Figure 4.1).

Koch was 37 years old before his *real* career began!

10

Koch at the Crossroads: From Lone Doctor to Group Leader

They were all crowded into a single large room with three windows. In front of each window there was a work table. Koch sat at the middle table and Gaffky and Loeffler sat on each side, separated from the master by nothing more than folding screens.

—Carl Salomonsen[1]

Koch's move to Berlin ushered in a new era in his research. Until now, he had been a lone worker, doing everything for himself, with only his wife Emmy as a sometime assistant. But now he was in a regular research institution in a major center of scientific and cultural activities, with the opportunity for assistants, co-workers, and colleagues.

It is sometimes the case that the scientist who works well alone experiences great difficulty working as part of a team. Koch appears not to have had this difficulty at all. Although he continued to do his own research in an effective manner for the rest of his life, he also had no problem directing the work of others. Soon he had a small group of enthusiastic and spirited co-workers who were ready to follow his lead.

His title in the *Kaiserliche Gesundheitsamt* was *Regierungsrath*, which translates literally as *Government Councillor*, a position in the German Civil Service. Although the *Amt* was just in the process of development, two laboratories were already in operation, one in hygiene under the direction of Gustav Wolffhügel, and the other in chemistry under the direction of Eugen Sell. Koch was in charge of the third (newly established) laboratory for bacteriological research.

When the *Amt* had begun to undertake its own research, the house at 57 Luisenstrasse (see Figure 9.2) had been remodeled into laboratories for chemistry and hygiene. When Koch arrived there was very little room to spare.[2] He was assigned a small room with a single window, which was quite in contrast to the rather elegant laboratories of the other two groups. However, Koch set to work anyway, unconcerned about the condition of his laboratory. Compared to Wollstein, anything must have been an improvement!

The Kochs rented a five-room apartment in the Chausseestrasse, not far from the laboratory in Luisenstrasse. From this apartment, Koch could easily walk to the laboratory in a few minutes.

Soon Koch had his first assistants, Georg Gaffky (1850–1918) and Friedrich Loeffler (1852–1915), both of whom were to become major figures in bacteriology (Figure 10.1). Loeffler had been transferred from the General Headquarters of the Army to the *Gesundheitsamt* a year earlier, and had worked as an assistant in the hygiene and chemistry laboratories during that time. When Koch arrived on 10 July 1880, Loeffler requested that he be allowed to work under Koch.[3] Loeffler made a number of important contributions, both in Koch's laboratory and later. He is best known for his discovery of the causal agent of

Figure 10.1   *Friedrich Loeffler and Robert Koch.*

diphtheria, *Corynebacterium diphtheriae*, and for his proof that the causal agent of foot-and-mouth disease was filterable, the first animal virus to be so characterized.

Gaffky, Koch's other assistant (see Figure 15.1), had been an assistant physician in the Royal Prussian Medical Corps and was assigned to the *Kaiserliche Gesundheitsamt* specifically to work with Koch.[4]

Thus, the three of them, Koch, Loeffler, and Gaffky, set to work in the small room on the upper floor of 57 Luisenstrasse. Some of the most important work of Koch's life was done in these crowded quarters. Later, he was able to move to a somewhat larger room that had three windows, but was still not especially spacious. Carl Salomonsen reported on the conditions at this time:

> I was the first foreigner to work in Koch's laboratory at the Imperial Health Office. At the time I was there, they were all crowded into a single large room with three windows. In front of each window there was a work table. Koch sat at the middle table and Gaffky and Loeffler sat on each side, separated from the master by nothing more than folding screens.[5]

Why were the windows so important? These were the days before electrical current and microscopy was often done using outdoor illumination. Later Koch had a laboratory all to himself next to that of his assistants, and an additional small room for photography.

Soon there were other co-workers: the bacteriologist Ferdinand Hueppe (1852–1938), later Professor of Hygiene at the German University of Prague; the chemist Bernhard Proskauer (1851–1915), who began in the chemical laboratory and transferred to Koch's group (where he remained for many years); service workers and secretarial help. Later, Gustav Wolffhügel, the leader of the hygiene laboratory, began to collaborate with Koch on certain projects. In a short time, Koch was surrounded by an enthusiastic group of dedicated workers. Thus, his position at the *Amt* brought him not only more space and time for research than he had at Wollstein, but interaction with intelligent colleagues. Koch matured rapidly in this new milieu, moving readily from lone scientist in Wollstein to group leader in Berlin.

How did Koch manage to pull together this marvellous and effective group so rapidly? In those early years, not through force of personality or an insistence on obedience, but through example. No one worked harder than Koch, and his example drove the others of his group forward with him. Sitting at his microscope, surrounded by his colleagues, daily making new and exciting discoveries: this was the way Koch led his group. We can be amazed at how rapidly significant research was carried out. By 1881, the first volume of the *Mittheilungen aus dem Kaiserlichen Gesundheitsamte* (Reports of the Imperial Health Office) was published[6]. This weighty book contained the major paper by Koch on the plate technique for isolating pure cultures (see Chapter 11), plus another important paper by Koch on disinfection, which set the way for further work in this important area. Koch also had a paper on anthrax which was partly an attack on Pasteur (see Chapter 16). In addition, there were papers by Loeffler, Gaffky, Wolffhügel, Hüppe, and Sell. In all, an impressive first year for the Imperial Health Office! We discuss details of some of Koch's work in the Imperial Health Office in the following few chapters.

11

Simple Gifts: The Plate Technique

The pure culture is the foundation of all research on infectious disease.

—ROBERT KOCH[1]

Perhaps Koch's greatest contribution to the development of bacteriology and microbiology as independent sciences was his introduction of a pure culture technique using solid or semi-solid media—soon known throughout the world as "Koch's plate technique" (*Plattenverfahren*).

The pure culture technique before Koch

Pure cultures in the sense we know them today were not obtained by Louis Pasteur or members of his school. Pasteur grew bacteria in transparent liquid media. When growth occurred, as evidenced by the development of turbidity in the culture tube, a minute quantity of the culture was inoculated into a fresh medium, and so on in series. By means of serial transfer, Pasteur assumed that a "pure" culture of one type of microorganism would ultimately result. Purity was ascertained primarily by microscopical examination. It was possible in many cases to select a medium that was the most appropriate for a single type of organism and hence obtain some degree of "purity".[2] However, if a

94

pure culture was obtained with such procedures, it was just fortuitous. Pasteur's cultures were equivalent to what today would be called "enrichment cultures".

Joseph Lister was the first to obtain a pure culture in liquid medium using a limiting dilution method. Lister's work, published in 1878,[3] was motivated by his desire to show that a single type of bacterium was responsible for a single disease entity. However, rather than studying an infectious disease, he studied the souring of milk by the bacterium that he called *Bacterium lactis*.

> I selected the lactic fermentation . . . because the effects which it produces in milk are extremely striking and readily recognized—the solidification which takes place being obvious at a glance, and the souring as shown by test paper being also a very conspicuous change.[4]

Using rather quaint equipment (his culture vessel was a liqueur glass), Lister made a series of successive dilutions of a lactic culture, choosing the tube from the highest dilution that soured as the source of inoculum for his subsequent cultures. This method led to cultures that were almost certainly pure. More important than Lister's method, however, was his clear enunciation of the *importance* of pure cultures for studies in pathology.

Since Lister was one of Koch's champions in England (it was Lister who had arranged for the English translation of Koch's 1878 book, see Chapter 8), Koch was certainly aware of Lister's work. He was also aware of the importance of the pure culture for studies on infectious disease. However, until he went to Berlin, methods for obtaining pure cultures in a consistent and reproducible manner eluded him. One of the first tasks he set himself in his new position was the perfection of all the methods needed for studying pathogenic microorganisms. It is not accidental that the first paper in the first volume of contributions from the Imperial Health Office is Koch's paper on methods for studying pathogens.[5] If one were to choose a single paper as most significant for the rise of microbiology, this would be it. Koch presents a method for isolating pure cultures that is so simple, reproducible, and understandable that it can be performed by almost anyone. During the two decades following its publication, the development of this method led to the isolation and characterization of the causal organisms of almost all of the major bacterial diseases which affected humans (see Chapter 22).

Background of the plate technique.

The basis of the plate technique is the development of isolated colonies on solid or semi-solid surfaces. The development of pigmented colonies on the cut surfaces of incubated potatoes was first reported in Breslau in 1875 by Joseph Schroeter (1835–1894), a student of Ferdinand Cohn's.[6] Schroeter studied such pigmented organisms as *Bacteridium prodigiosum* (now called *Serratia marcescens*) and *Bacteridium violaceum* (now called *Chromobacterium violaceum*), pointing out that the color of the pigment was a constant characteristic of the bacterium. He observed that pigment was formed only when the bacterium was growing on the surface of a nutrient substance and concluded that air was necessary for pigment formation. He did a number of studies on the effects of environmental conditions on pigment formation and did simple extraction studies to obtain some idea of the chemistry of the pigments. But the importance of Schroeter's work rested on his simple observation that each organism formed a characteristic pigment and that the ability to form this pigment remained upon transfer to a new medium. Schroeter used not only potatoes, but solid media made from starch paste, egg albumin, bread, and meat. It is clear from reading Schroeter's paper that he definitely obtained pure cultures of a wide variety of pigmented bacteria. Koch, a frequent visitor to Cohn's laboratory, was certainly familiar with Schroeter's work, although he does not cite it.

Another important predecessor of Koch was Oscar Brefeld (1839–1925), a German mycologist who made many important contributions to the understanding of the fungi. In 1875, Brefeld laid down precisely the principles which must be followed for obtaining pure cultures.[7] These principles were:

1. The inoculation of the medium should be made from a single fungal spore.

2. The medium should be clear and transparent and should yield optimal growth of the organism.

3. The culture should be kept completely protected from external contamination throughout its existence.

Brefeld's method of inoculating a medium with a single spore was completely suitable for the study of large microorganisms such as fungi but was less useful with the much smaller bacteria. However, his writings had wide influence on the bacteriologists of the day, and Koch certainly built his technique around Brefeld's principles.

Koch's 1881 paper

The paper which Koch published in the 1881 *Mittheilungen* on methods for the study of pathogenic organisms became the "Bible of Bacteriology".[8] The paper not only includes the major section on pure cultures to be discussed here, but also an extensive section on photomicroscopy of bacteria, with the first published photomicrographs of bacteria in diseased tissues (Figure 11.1). Photomicroscopy of bacteria in infected tissues was an accomplishment that had eluded Koch in his 1878 work, as discussed in Chapter 8.

Koch began the section on pure cultures with this solidly based statement:

> The pure culture is the foundation for all research on infectious diseases.

He then continued:

> The most important procedures that have been developed for the manipulation of pure cultures can be summarized as follows.
>
> A sterilized container is used which has been closed with mold-proof sterilized cotton,[9] and this is filled with a sterilized nutrient liquid of the proper sort. Then this is inoculated with material containing the microorganism which is wanted in pure culture. After reasonable growth has taken place, a sterile instrument is used to transfer a little of this to a second container. This process may be repeated a number of times. . . .
>
> Naturally in this procedure one has to make several assumptions, of which the first is that the culture vessel is really sterile. How lightly this sterilization has occasionally been treated can be seen from the controversy between Pasteur and Bastian on spontaneous generation, and the well-known question of the former to the latter: "Flambez-vous vos vases avant de vous en servir?", which Bastian had to answer in the negative. . . . [Among other assumptions, the most important is] that the substance used as inoculum contains no other microorganism than the one desired. Even a slight contamination of the inoculum with another species which is faster growing than the organism desired will prevent anyone from ever obtaining a pure culture. . . . All in all, the present situation with regard to pure culture techniques is quite disappointing. Anyone who has cultured microorganisms in the ways currently in vogue will have found how difficult it is to avoid completely all of the sources of error that I have indicated. No one following current methods can complain if his results

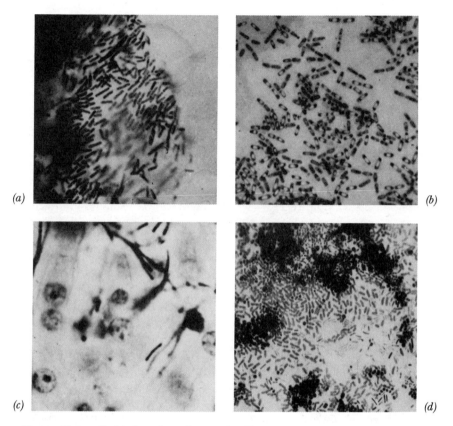

Figure 11.1   *Koch's first photomicrographs of bacteria in diseased tissue. (a) Section through kidney epithelium of a guinea pig kidney dying of an infection with* Bacillus anthracis. *(b) Spore formation by bacillus from a cadaver. (c) Section through the liver of a rabbit infected with anthrax. (d) Bacteria from blood that had putrefied for several days, as an example of the diversity of bacteria in putrefying situations (as opposed to the uniformity of bacteria in diseased situations).*

are not accepted as fact by his fellow workers. What has been said above should be heeded by the Pasteur school[10] in its noteworthy but blindly zealous researches, since this renders it doubtful that they have obtained in pure culture the organisms of rabies, sheep pox, tuberculosis, and so forth. . . .

The present methods seem to me to offer no hope for a significant improvement. . . . Therefore, I have rejected completely all of the current principles of pure culture technique and have adopted an entirely new way.[11]

A forthright statement, indeed! In his paper, Koch then turned to an explanation of the rationale for the plate technique that he had developed. First, he discussed the development of colonies on the surfaces of cut potatoes, following Schroeter's lead (see above).

> What can we conclude from these observations of colonies developing on potatoes? . . . Most often each colony is a pure culture and remains a pure culture until it enlarges to the point that it touches its neighbors. If instead of the potato, a liquid medium of the same surface area were exposed to the air, then undoubtedly the same number and the same kinds of germs would develop as on the potato, but the development of these germs in the liquid would be different . . . Some of the organisms would sink to the bottom of the liquid, while others would rise to the top. Some of the organisms which would have found places on the potato to grow undisturbed would be choked by the development of other more luxuriantly growing organisms and would never grow. In short, the whole liquid would reveal under the microscope . . . a tangled mixture of different shapes and sizes, which no one would mistake for a pure culture. What is the fundamental difference between the nutrient substratum which the potato offers and that offered by the liquid medium? It is only that the potato is solid and prevents the various species, even if they are motile, from becoming mixed . . .
>
> How then can we make use of the advantages which a solid nutrient medium offers for the practice of the pure culture? A number of colonies which had developed spontaneously on a boiled potato were spread out on other similar potato slices and incubated in the moist chamber. Within one or two days a heavy growth of the seeded microorganism develops, with exactly the same characterstics as those from the original droplet. . . . All of them grow quite quickly from very small colonies of the original potato when transferred to other potatoes and appear to be perfectly pure cultures. . . .
>
> Here therefore is a very simple method for the production of perfect pure cultures, at least for those organisms which can grow on boiled potato . . . However, pathogenic bacteria cannot be cultured on potato.
>
> But the principle had been found, and it was only necessary to devise conditions which could be used in all cases.

Koch now discusses the kinds of media which pathogenic bacteria can grow on, such as nutrient broths and boullions. Instead of trying to find a solid medium which would support pathogens, he *took a medium which supported the growth of pathogens and made it solid*, by adding gelatin.

> The mixture of nutrient liquid and gelatin, which I will call nutrient gelatin for short . . .

The details of the preparation and use of nutrient gelatin, so carefully given in Koch's paper and so important for the research worker, need not concern us here.

> I have carried pathogenic and nonpathogenic organisms over a long series of transfers on boiled potato or nutrient gelatin, without ever once observing any noticeable changes in their characteristics. They maintain their morphological as well as their physiological characteristics . . . without change through months of growth as pure cultures.

It is easy to see how this paper of Koch's became the "Bible of Bacteriology" (Figure 11.2). In a few pages, Koch laid out clearly the essential problem of bacteriological research and provided a solution— a solution so general and widely applicable that it changed for all time the field of bacteriology. Realizing the importance of obtaining pure cultures, Koch took Schroeter's simple observation of the growth of bacterial colonies on potato slices and adapted it to the field of medical bacteriology. The initial challenge arose because many of the pathogenic bacteria Koch was interested in would not grow on potatoes but would grow on nutrient broth media. The solution, therefore, was to convert the nutrient broth into a solid medium. Once the value of this approach had been realized, the whole procedure was generalized. Then it could be applied not only to the growth of pathogenic bacteria but to all kinds of microorganisms. Koch's work described in this paper was an amazing tour de force, one rarely duplicated in the scientific world.

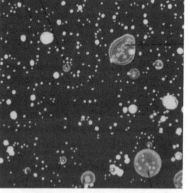

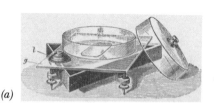

(a)

(b)

Figure 11.2   *The Koch plate technique, as illustrated in a textbook widely available in Koch's time.[16] (a) Preparing a plate. Note the careful levelling technique. (b) Photograph of colonies. In the early years of the plate technique, colony densities were too high for accurate counting. A knowledge of statistics of colony counting did not develop until early in the 20th century.*

Early uses of the plate technique

Koch realized immediately that the plate technique had many uses besides its value for the isolation of pure cultures. Most important, the technique could be used to assess the numbers and kinds of microorganisms found in various environmental samples, such as air, water, soil, food, manufactured objects, etc.

The importance of such studies was enormous. Not only was the assessment of the microbial content of environmental samples important for public health research, it aided greatly in the improvement of bacteriological technique. Once a quantitative understanding of the microbial populations of habitats such as air and water could be obtained, the development of improved techniques for the laboratory study of microbes became much easier.

Koch was quick to communicate his methods to others. He set up demonstrations in the laboratories of the Imperial Health Office and details of such demonstrations were published in widely available journals.[12]

The introduction of agar

Nutrient gelatin was a marvellous culture medium for isolation and study of pure cultures of bacteria, but it had several drawbacks. The most important drawback is that gelatin does not remain solid at body temperature. Thus, one had to incubate cultures at temperatures below their growth optima. In initial studies, this presented no major obstacle, and there are certainly few human pathogens that will not grow at all at room temperature. However, for detailed studies, and for incubation during the summer when room temperature was too high for gelatin to remain solid (these were the days before refrigeration and air conditioning), something better was needed. This "something" was agar (or agar-agar, as it was called in those days).

Agar is a polysaccharide derived from seaweeds of the Rhodophyta. It was used widely in the 19th century, especially in tropical countries, as a gelling agent and had found its way into Europe via trade of the European countries with their tropical colonies. The first use of agar as a solidfying agent for bacteriological culture media was made by Walther Hesse (1846–1911), an associate of Koch's (Figure 11.3). The story of the introduction of agar into bacteriology has been told in detail[13] and should be mentioned briefly here.

Figure 11.3   *The developers of agar, Fannie and Walther Hesse.*

Walther Hesse was a physician who worked for half a year in 1881/82 in Koch's laboratory, carrying out quantitative studies on the micro-organisms which were found in air. Hesse was a district physician in Saxony, and became a student in Koch's laboratory in order to learn the new discipline of bacteriology. He eventually published his work on bacteria in the air in Volume II of the *Mittheilungen aus dem kaiserlichen Gesundheitsamt*, but most of the work was done in his home after he left Koch's laboratory. The actual suggestion to use agar instead of gelatin was made by Hesse's wife, Fannie Eilshemius Hesse, who was not only Hesse's wife but also his technician and artist illustrator. (Hesse's 1884 paper was illustrated with three chromolithographs made by Fannie Hesse.) Fannie Hesse was actually born in the United States of German parents and had returned in 1874 to Germany, where she met and married Walther Hesse.

The counting of bacteria in the air was done by Hesse using a special apparatus that sucked air through tubes lined with gelatin, and the microbial content was assessed by incubating the tubes and counting the colonies which developed. However, in the summer, the gelatin melted during incubation, ruining many experiments. Hesse thus decided to seek a new solidifying agent. Fannie Hesse had used agar for years in the preparation of fruit jellies, following a recipe which she had received from her mother who had in turn learned of it from some Dutch friends who had lived in Java. Hesse tried agar in his nutrient tubes and found that it worked much better than gelatin. He wrote to Robert Koch about his discovery, and Koch rapidly adapted agar to his own studies. In his short paper on tuberculosis published in 1882 (see Chapter 14), Koch mentioned using agar as a solidifying agent. No formal paper on the use of agar was ever published.

The Petri plate

One final extension of the plate technique should be mentioned here: the development by Richard J. Petri (1852–1921)[14] of a special plate for agar or gelatin culture. In Koch's original method, flat glass plates were used (see Figure 11.2) which were layered with nutrient gelatin or agar on a special horizontal pouring apparatus. The poured plates were then placed on small glass shelves under a large bell jar for incubation. The glass plates and the bell jar were cumbersome, and the pouring apparatus involved a complex arrangement.

Petri's enhancement, which turned out to be amazingly useful, was the development of the special glass plate which bears his name. The plates that Petri first used were flat double sided dishes of 10–11 cm in diameter and 1–1.5 cm in height. The dishes were sterilized separately from the medium by dry heat, and after cooling, the nutrient agar was poured in. The medium solidified in a few hours and the plates could then be used. Petri noted that such plates dried out slowly, could be easily examined directly under the microscope, and colonies could be readily seen through the bottom of the dish. He especially noted the value of these plates for carrying out quantitative counts of bacteria.

Final words

The far-reaching implications of the Koch plate technique are obvious to all bacteriologists. Perhaps no technique has had such an important influence on the development of the field. When we contemplate this miracle now, we might wonder why the plate technique had not been thought of earlier. Certainly one of the major reasons was that earlier workers lacked the will to develop new techniques. As long as one had doubts about the germ theory of disease, there was little motivation for thinking up new techniques. But Koch was strongly committed, and he realized the importance of the pure culture. Building on Schroeter's pioneering work, but generalizing it completely, Koch rapidly perfected this method. And it was so simple to use that anyone could learn it. Even the French, who found themselves at odds with much of Koch's work, acknowledged the significance of Koch's plate technique. As Pasteur's closest associate Émile Roux said:

> Culture in solid media is very useful because it permits us to observe the form of the colonies and because it readily permits the separation of diverse organisms.[15]

Soon workers from all over the world were flocking to Koch's door to learn his techniques. In Chapter 13, we describe Koch's trip to London, where he introduced the plate technique to Lister and Pasteur, certainly a memorable occasion in the history of bacteriology.

12

Sterilization, Disinfection, and other Techniques

Robert Koch initiated a new era by comparing the action of a very large number of reputed antiseptics on certain bacteria. [Using] pure cultures of bacteria Koch found that Lister's carbolic acid had no disinfecting power at all.

—William Bulloch[1]

One of the main missions of the Imperial Health Office was public health, and Koch and his associates set to work enthusiastically to place various public health practices on a rational basis. Among the most important tasks was the development of methods for sterilization and disinfection. Proper procedures would be of major value not only in the laboratory study of microorganisms, but in surgical practice and in the control of the spread of infectious disease as well.

An important consideration in this discussion is the distinction between sterilization, which involves the complete killing of both spores and vegetative cells, and disinfection, in which vegetative cells are killed but spores are not necessarily all killed. In the early days of bacteriology, especially before quantitative methods were perfected, this distinction was often not clear.

Chemical disinfection

Lister had used chemical treatment of wounds as the basis of the antiseptic system of surgery which he first described publicly in 1868.

However, he had made no systematic study of chemicals appropriate for his method. He used carbolic acid because its antiseptic power had been described by earlier workers. The state of surgery before Lister introduced his method was so bad that almost any toxic chemical that did not kill the patient would have been found to be effective. However, Koch quickly showed that carbolic acid was poorly effective and carried out a systematic study to find a better agent. His paper, completed in April 1881 and published in the same volume of the *Mittheilungen aus dem kaiserlichen Gesundheitsamt* in which the paper on the plate technique was published, was a monument of clarity and logic.[2]

Up to Koch's time, the efficacy of chemical disinfectants had been determined by testing their ability to prevent putrefaction of one or another organic material. Putrescible liquids, such as infusions, were allowed to incubate in the presence of the disinfectant and microscopic observations were made for microbial growth. Later, when cultivation of bacteria became more common in the laboratory, workers transferred a small quantity of the initially incubated material into a suitable liquid culture medium and incubated further, observing for the development of bacterial turbidity. If turbidity developed, then it was inferred that the disinfectant had been ineffective, whereas if the medium remained clear, disinfection was considered to be complete.

Koch made several major criticisms of these procedures:

> In the first place, the experiment is made with an uncertain mixture of many species of bacteria, and no attempt is made to determine which varieties of bacteria, if any, are affected. Second, it is also a matter of chance whether spores are present. . . . Third, it is difficult to tell from the turbidity which develops whether the bacteria which grow were those present in the initial material, or were accidental contaminants from the air.
>
> I have, therefore, abandoned this method and have taken a completely new approach. I have used pure cultures of species of bacteria which are known to be important in putrefaction or disease. Solid culture media are employed instead of liquid, so that contamination can be ascertained. Small quantities of appropriate pure cultures are exposed to the action of disinfectants . . . and the resulting growth, if any, is compared with that occurring in control experiments, which are made in every instance. The disinfecting power of the agent against spores can also be determined independently.

The logic of Koch's approach is so obvious that one might wonder why it had not been done earlier. However, it required the availability of consistent cultivation methods on solid media and an understanding of the variable resistances of spores and vegetative cells. The introduction of solid media for antimicrobial testing was seminal: over 100 years later this is still the method of choice for the study of the action of antibiotics and disinfectants.

One of the major experimental problems in disinfection studies is the carry-over of the disinfectant chemical from the initial incubation mixture to the culture medium in which viability is to be determined. Koch was aware of this problem and attempted to avoid it by using as small a sample of the original incubation mixture as possible and transferring it to a relatively large volume of culture medium. In this way, the disinfectant was diluted out. When he tested spores, he used another procedure which was even more advantageous. He soaked short pieces of silk thread with anthrax spores and allowed them to dry. (Since spores are resistant to drying, they could be treated in this way, whereas vegetative cells could not be so tested.) Then, after the disinfection period was complete, the threads were carefully removed, the disinfectant washed off by means of sterilized distilled water, and the threads placed on a culture medium for incubation (or, in some cases, inserted into animals to determine if the spores were still pathogenic).

Koch recognized that disinfection was a quantitative event, and that not all organisms in the disinfection mixture would be killed simultaneously. He also realized that the time required for complete disinfection depended upon the concentration of chemical used. He emphasized the importance, therefore, of determining the time/concentration relationship of killing.

One of the most important principles enunciated by Koch was that *inhibition of growth* and *killing* were two separate activities. Carbolic acid and certain other chemicals were quite capable of inhibiting the growth of bacterial spores, but did not kill them.

Using his carefully constructed methods, Koch proceeded to test Lister's carbolic acid in detail. He readily showed that although carbolic acid was capable of inhibiting *growth*, it had no *disinfecting power* at all against anthrax spores. This was true no matter what kind of environmental conditions were used during the disinfection experiment. Two other disinfectants that had been widely used, sulphurous acid and zinc chloride, were also found to be ineffective.

Having shown that the three most widely used disinfectants were useless, Koch then proceeded in a systematic way to test a wide variety of chemical agents. For these comparative studies, anthrax spores dried upon silk threads were used. Anthrax was a logical choice, since Koch had so much experience with the anthrax bacillus. He incubated the infected threads in the chemicals for various periods of time (measured in days), then removed threads at intervals and determined viability by incubation in culture medium. The shortest time needed to completely kill the spores was noted, permitting quantitative assessment of the various chemical agents.

Koch found many chemicals to be completely ineffective in this (rather stringent) test, despite the beliefs of earlier workers that particular chemicals were good disinfectants:

> It is surprising that a number of substances which are generally assumed to be destructive of organic life have absolutely no effect on the spores of the anthrax bacillus. I mention in this regard hydrochloric acid (2% concentration), sulfuric acid (1% concentration), and saturated solutions of calcium and sodium chloride. Furthermore, metal salts and such chemicals as boric acid, potassium chlorate, benzoic acid, sodium benzoate, and quinine have only a slight effect. . . . Out of a long list of substances, we have left only mercuric chloride and the halogens chlorine, bromine, and iodine.

Koch noted that in places where gaseous disinfectants such as the halogens were inappropriate, the only effective disinfectant was mercuric chloride. As Koch noted, this compound had rarely been employed by earlier workers because of its poisonous properties. However, it was the only known disinfectant that could destroy spores in the dry state, and at a high dilution (1 to 1000 or even 1 to 5000).

> The sole drawback to its use is its poisonous nature; but owing to its rapidity of action only a quarter or half an hour is needed for disinfection, and the reagent can then be removed by thorough washing with water. Minute traces will doubtless remain after washing, but would be absolutely harmless.

Koch's demonstration that carbolic acid was ineffective as a disinfectant was quickly accepted by Lister, who then began to adopt different chemicals and improved methods for his antiseptic surgery[3]. Koch's studies also provided the groundwork for subsequent research on chemical disinfection, most importantly that of Krönig and Paul on which all modern work is based.[4]

Koch's work on chemical sterilization was important not only because it led to the discovery of new and more effective antiseptics and disinfectants, but because it provided a precise and reproducible *method* for studying the whole disinfection process. His work formed the essential foundation for all later work in this important area.

Heat sterilization

The necessity for the sterilization of culture apparatus and media was recognized early by most workers. The use of heat for this purpose arose initially from the extensive work on spontaneous generation by Pasteur. However, Pasteur did not recognize that entities such as heat resistant spores existed. This discovery rested with Tyndall and Ferdinand Cohn (see discussion in Chapter 6). Tyndall introduced the technique of fractional sterilization, a method which permitted the killing of heat-resistant endospores without use of temperatures higher than boiling. However, Tyndall's method was time-consuming and cumbersome. The practical application of heat sterilization to bacteriological technique was worked out by Robert Koch and his associates at the Imperial Health Office. In the first volume of the *Mittheilungen*, papers were published on the use of heat for sterilization. Koch and Gustav Wolffhügel contributed the paper on the value of hot air as a sterilizing agent[5] and Koch, Gaffky, and Loeffler determined the limitations of steam sterilization[6].

Sterilization by hot air

For the work on hot air, Koch and Wolffhügel used various pure cultures of microbes, including both sporeforming and nonsporeforming organisms. The procedures for these experiments with heat could be simpler than those for chemical disinfection, since the agent did not have to be removed after treatment. After cooling, the cultures were transferred to suitable media and their viabilities determined by incubation.

For their heating experiments, Koch and Wolffhügel used chambers which were heated by means of coils of copper tubing lining the interior, through which compressed steam was passed. The temperature of each chamber was regulated via regulation of the steam pressure[7]. The maximum pressure allowed by the steam boiler was 5.5 to 6 atmospheres, which permitted a temperature of 140°C within the chamber. Although this procedure permitted testing a number of organisms at once, it had the difficulty that the temperature distribution within the chamber was quite unequal.[8]

Using these procedures, it was shown that nonsporeforming bacteria were destroyed in 1.5 hours at a temperature slightly exceeding 100°C, whereas sporeforming bacteria required 3 hours at 140°C. The use of Tyndall's fractional sterilization would permit use of lower temperatures, but Koch and Wolffhügel pointed out that Tyndall's method depended upon the ability of the spores to actually germinate and grow during the interval between each heating period. Since many objects that must be sterilized would not support microbial growth, Tyndall's method would not always work.

> Almost all objects requiring disinfection are merely carriers of germs, and not culture media, so that if the spores clinging to such objects withstand the first heating they still remain as spores and are unaffected by subsequent heatings.

Koch and Wolffhügel also studied the penetration of dry heat into articles of larger size, such as pillows, and showed that due to their insulating properties, even 3 to 4 hours exposure at 140°C was insufficient. On the other hand, if longer exposures were used, the fabric would be damaged. These ideas naturally led to a discussion of sterilization with steam, the subject of the paper by Koch, Gaffky, and Loeffler.[6]

Steam sterilization

Sterilization by heat under pressure had been the cornerstone of the food canning industry for many years, the practice first having been introduced in France by Nicholas Appert (1750–1841). Pasteur had adapted Appert's method to wine, using mild heat to destroy spoilage organisms so that the keeping qualities of the wine were extended (the origin of the pasteurization process). However, there had been no systematic study of the use of steam sterilization for bacteriological purposes.

For their work on steam sterilization, Koch, Gaffky, and Loeffler used steam at both atmospheric pressure (Figure 12.1*a*) and under increased pressure. For the pressure experiments, a closed iron vessel was used which was 20 cm in diameter by 40 cm in height, filled to one-fifth its capacity with water, and heated by gas burners (Figure 12.1*b*).

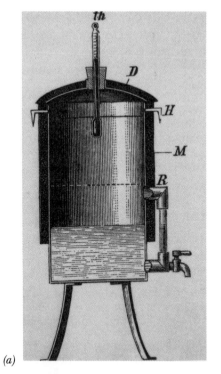

(a)

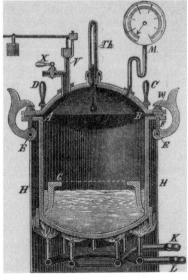

(b)

(c)

Figure 12.1   *Apparatus for heat steriliza-tion. (a) Sterilization in steam. (b) Pressure ster-ilization. (c) One of the early steam sterilizers used in Koch's laboratory. From a museum ex-hibit.*

Although they did not measure the pressure, they determined that flasks of cold water standing in steam which itself reached temperatures of 120°C within 15 minutes had only reached 102°C in 30 minutes. They realized immediately the importance of determining the temperature not in the steam chamber but in the test sample itself.

> Hence it must not be assumed that bodies exposed to steam readily attain the temperature of the steam. We must therefore accept with the greatest reservation reports of cases in which hay infusions have been said to have been heated by steam to temperatures of 100°C for hours without being sterilized.

Koch and his co-workers then studied the rate at which bulky objects attained equilibrium with the temperature of their surroundings and found that effective heating took a long period of time.

Concerning spores themselves, Koch and his colleagues determined the time/temperature relationship for complete sterilization of both anthrax spores and spores of a soil bacillus. The anthrax spores were killed by ten minutes at 95°C whereas the spores of the soil bacillus required ten minutes at 105°C. Although steam under pressure was more rapidly effective than steam at 100°C, Koch and his colleagues preferred the latter condition whenever possible because it was easier to arrange for heating at atmospheric pressure.

> These results leave no room for doubt as to the form of disinfection by heat which should be adopted in the future. The hot air apparatus is complicated and costly, and is untrustworthy when the objects to be disinfected are at all bulky, or folded, or wet. Disinfection by steam under pressure at temperatures above 100°C is open to the same objections, though to a lesser degree. In every respect, exposure to flowing steam at 100°C is far more satisfactory. It is more certain, more simple, more rapid, more economical both in original cost and expense of working, and involves less injury to the articles to be disinfected.

Experience would eventually show that this conclusion was erroneous. Heating in steam under pressure is necessary for the effective sterilization of most objects. As the technology for steam sterilizers developed, the difficulty of sterilizing under pressure diminished. However, the contribution of Koch and his co-workers was major. They showed that moist heat is much more effective than dry heat, and they showed the importance of measuring the temperature at the center of the object to be sterilized. For bulky objects, it is essential that the time used be sufficiently long that penetration of heat to the center occurs.

Eventually, heat sterilization procedures, as used in either the bacteriological laboratory or the hospital, were refined so that there was a considerable margin of safety. All subsequent work built on the careful quantitative studies of Koch and his associates.

Final words

The work of Koch and his associates at the Imperial Health Office played an exceedingly important role in placing the whole study of sterilization on a firm footing. The methods which Koch developed have come down to us today, refined of course, but basically the same. By the use of these sterilization methods and the plate technique described in the last chapter, bacteriology now had the tools for a concerted attack on the causal agents of most infectious diseases. And it was in this direction that Koch turned, beginning the important work which was to occupy him for most of the rest of his scientific life.

13

The London Meeting: Koch, Lister, and Pasteur

C'est un grand progrès, Monsieur

—Louis Pasteur[1]

In the summer of 1881, Robert Koch went as a delegate of the German government to London to attend the Seventh International Medical Congress. This important meeting, held from August 2 to 9, brought together most of the active medical researchers in the world. Although Koch did not present a paper, he made an important demonstration of the plate technique as well as other methods which he had developed.

As we have seen, Koch's work had come to the attention of Joseph Lister several years earlier, and Koch's 1878 book was translated into English at Lister's instigation (see Chapter 8). Now in London for the first time, Koch found himself surrounded by many interested workers who were following Lister's lead. Lister not only mentioned Koch in the speech he gave to the Pathology Section of the meeting, but also arranged for Koch's demonstration to be set up in Lister's own laboratory at King's College. We discussed in Chapter 11 the development of pure culture techniques and showed that Lister was one of the first to obtain an indubitably pure culture, using his limiting dilution method. It was Koch's demonstration of his plate technique that was to be a major event of the Congress. However, it seems evident from Lister's announcement of the forthcoming demonstration that he was not yet aware of Koch's plate technique, which had not been published at the time of the London meeting. Rather, Lister had in mind a demonstration of Koch's microscopic and photographic methods. Thus:

> I have the great satisfaction, gentlemen, of knowing that Dr. Koch is present among us, and also that, with infinite kindness, and very great trouble to himself, he is about to exhibit at King's College, to a limited number, his methods of procedure in actual operation. It is but to a limited number that these demonstrations can be made, because only a very few at a time can see them to advantage; but I have also the great satisfaction of knowing that Dr. Koch will exhibit this afternoon in this room by the magic lantern photographs of sections made by himself of various diseased tissues, illustrating the effects of micro-organisms. These photographs will be as convincing and as satisfactory as the actual demonstration of Dr. Koch's processes, because the pictures drawn by light are entirely free from those errors which can hardly fail to creep in as a consequence of mental bias when a sketch of these minute objects is made by the human hand. Permit me to return to Dr. Koch the thanks of this Section for his great kindness in this matter.[2]

Also present at the London Congress was Louis Pasteur, now at the height of his powers. Pasteur had made his famous demonstration of the anthrax vaccine at Pouilly le Fort earlier in the year and was just in the early stages of his work on attenutation of the causal agent of fowl cholera. At the London meeting, Pasteur was received everywhere with acclaim. He reported details of his fowl cholera studies, studies that were to lead him on the road to the major work of the last quarter of his life— the use of attenuation in the development of vaccines against infectious disease. Lister and Pasteur had been in frequent contact, since it was Pasteur's work on fermentation and spontaneous generation that had first led Lister to antiseptic surgery.

Unfortunately, the details of Koch's demonstrations have not been recorded in the transactions of the Congress, nor did Koch contribute a written report. According to Lister's biographer,[3] Koch's discussion at King's College was translated from German into English as his words were spoken by Dr. Frank Payne of St. Thomas's Hospital. The only written statement regarding the meeting was that of Lister, made some years later.[4] According to Lister, it was at this meeting that Koch first demonstrated his plate technique. This technique, of course, quickly superseded Lister's limiting dilution method, and it was understandable that Lister should have been excited about the demonstration of this method by Koch.

As we have noted, the illustrious Pasteur was present at the demonstration, and is said to have taken Koch's hand and exclaimed:

C'est un grand progrès, Monsieur.[5]

This was, indeed, a great triumph for Koch, as Pasteur could never forget that France had lost the Franco-German War of 1870. Indeed, as we will see in Chapter 16, Pasteur and Koch were soon to be in open conflict.

Only a few days after he returned from his London triumph, Koch began his work on tuberculosis, work that was to make his name known, not only throughout the scientific world, but to the general public as well.

14

World Fame: The Discovery of the Tubercle Bacillus

Koch was not optimistic about how his work would be received. On the way to the lecture, he predicted to me that it would take a year of hard battle before his discovery of the tubercle bacillus would be accepted by the medical profession. I, full of optimism and hope, predicted that his work would be quickly accepted.

—FRIEDRICH LOEFFLER[1]

It is one of the great scientific discoveries of the age.

—NEW YORK TIMES[2]

Less than two weeks after Robert Koch returned from his triumphant success in London, he began his research on the etiology of tuberculosis. It is fair to say that no single discovery in infectious disease has had more wide-reaching influence than Koch's discovery of the causal agent of tuberculosis. At the time Koch began his work, one-seventh of all reported deaths of human beings were ascribed to tuberculosis, and if one considered only the productive middle-age groups, one-third of the deaths were due to this dread disease.[3] And the disease was not limited to any socio-economic group, attacking equally rich and poor. In John Bunyan's famous phrase, tuberculosis was the "Captain of the Men of Death".[4]

The impressive thing about Koch's work on tuberculosis was not only its scientific brilliance, but the *speed* with which it was accomplished. He began the first experiments on 18 August 1881. And on 24 March 1882, less than 8 months later, he gave his famous and historic lecture on the tubercle bacillus to the Berlin Physiological Society, a lecture which was published only three weeks later.[5,6]

The Background on Tuberculosis

Tuberculosis had been recognized as a specific disease entity since antiquity.[7] There is evidence of tuberculosis lesions in the bones of Egyptian mummies. Although tuberculosis of the lungs does not seem to have been common in Egypt, *pulmonary tuberculosis* (also called *phthisis*) was well recognized by the Greeks, and extensive descriptions can be found in the writings of Hippocrates and others. Another major form of tuberculosis was subsequently recognized, *miliary tuberculosis*, in which the lesions are tiny nodules disseminated throughout the body. By the early 19th century, pathology was a well-developed discipline, but many pathologists held the view that miliary tuberculosis and phthisis were two distinct diseases, a view strongly (but incorrectly) supported by Rudolf Virchow. The French pathologist René Laennec held, on the other hand, that miliary tuberculosis and phthisis were two aspects of the same disease, and that tuberculosis was a morbid process that could occur in various parts of the body. According to Laennec, phthisis (called consumption in England) was tuberculosis of the lungs.

At the time Koch initiated his work, there was strong evidence that tuberculosis was a contagious disease. The French physician Jean Antoine Villemin (1827–1892) had showed in 1865 that tuberculosis was transmissible to experimental animals,[8] a discovery that was amply confirmed by Edwin Klebs, Julius Cohnheim, and Carl Salomonsen. But the suspected causal organism had never been seen, either in diseased tissues or in culture.

Koch's approach

When Koch began his work, he knew from Villemin of the value of the experimental animal for tuberculosis studies and he had access to extensive pathological material from the "phthisis ward" at the Berlin Charité Hospital. Koch's aim, from the beginning, was the demonstration of a parasite as the causal agent of tuberculosis. To this end, he employed all of the methods that he had so carefully developed over the previous six years: microscopy, staining of tissues, pure culture isolation, animal inoculation. These methods, which worked so well with anthrax and wound infections, also worked with tuberculosis, but not without great difficulty. The causal agent of tuberculosis was exceedingly difficult to stain and visualize in diseased tissues, and it grew so slowly in culture that extreme patience and a *strong faith* in its existence were necessary. And it was, indeed, *because* of Koch's faith in the parasitic nature of tuberculosis that he was able to succeed where others had failed.

Staining the tubercle bacillus

As is now well known, *Mycobacterium tuberculosis*, the tubercle bacillus, is very difficult to stain with conventional bacteriological stains because of its extremely waxy nature. One of Koch's main contributions was the discovery of a *method* for staining the tubercle bacillus. How did he do it? We have some insight into this phase of Koch's work from the remembrances of Loeffler, one of Koch's co-workers:[9]

> Numerous inoculations of guinea pigs with tuberculous material from various sources gave Koch the same clinical and pathological-anatomical picture and convinced him that he was dealing with a characteristic living agent. Driven forward by this conclusion, Koch set to work to demonstrate the presence of this agent in diseased material. He turned therefore to freshly developed tubercles, which always appeared first upon inoculation. He removed some of this material, streaked it out on cover glasses, and stained it with various dyes, using procedures that we had long used for other bacteria in diseased processes. Ehrlich's methylene blue, which Koch had used for a long time [see Chapter 8], was his first choice. In such stained preparations, Koch saw very tiny thin rods, about twice as long as wide. He found these rods only in preparations from tuberculous material, and not in controls. Were these rods the sought-for agent? This question was not easy to answer. First, Koch decided to obtain photomicrographs of these bacilli, in order to obtain an objective view of the organism, as he had always done in his earlier work. However, at that time the photographic technique was not very well developed and obtaining good pictures of stained material presented numerous difficulties. The technique that Koch found best was to counterstain his preparations with the brown dye *vesuvin* [Bismarck Brown] and then to photograph these brown-stained preparations with blue light. The brown-staining parts of the preparation absorbed the blue light and appeared dark on the photographic negative, whereas the blue-stained bacteria appeared to be bright and transparent. Although Koch counterstained the preparation with vesuvin to increase the photographic contrast of the blue-stained rods, when he examined these preparations before photography, he was surprised to discover that in the totally brown background the small rods had retained their blue color. Within the brown background they were now easily visible in large numbers! Extensive further experiments convinced Koch that he had found a new, very valuable method for differ-

entiating the bacteria in tuberculous material from other bacteria, and that it would be possible, with this method alone, to distinguish the tubercle bacillus from thousands of other bacteria. How well I remember that moment when Robert Koch showed me, for the first time, such a brown-stained preparation with tiny, but clearly visible, blue-staining rods. However, the brilliant research talent of Robert Koch was soon to be illuminated in even greater degree. After he had used his new technique to demonstrate the presence of the characteristic rods in all possible tuberculous tissues and fluids, he considered it necessary to repeat the whole experiment with freshly prepared dyes. But when he examined his new preparations, which had been stained for 24 hours in fresh methylene blue solution and counterstained with vesuvin, he sought in vain under the microscope for the blue-staining rods. However, using the same samples he could easily demonstrate the presence of the rods using the dye that had been prepared earlier. Therefore, something must have happened to the old dye solution that made it suitable for staining the tubercle bacillus. What? Koch concluded that the dye solution must have absorbed from the air something that made it suitable for the staining technique. One of the most common constituents in the air of a laboratory is ammonia, and Koch quickly concluded that the methylene blue solution had absorbed small amounts of ammonia from the air during its long stay in the laboratory. He then added a small amount of ammonia to his freshly prepared methylene blue solution and found that it now worked satisfactorily for staining the tubercle bacillus. Since ammonia is a strong alkali, the methylene blue solution could be made effective by adding any alkali, such as sodium hydroxide or potassium hydroxide. Through extensive and careful experiments, Koch determined the optimum concentration of alkali and the proper staining procedure. Koch mentioned nothing about this in his first lecture, nor in the paper that was published soon after. It was Paul Ehrlich, in a paper given at a meeting of the *Verein für innere Medizin* in Berlin on 1 May 1882, who improved Koch's staining procedure by using aniline instead of ammonia, and fuchsin instead of methylene blue.[10]

The staining of the tubercle bacillus, as prosaic as it might seem, was the key which opened the door to the mystery of this disease. Once the organism could be readily demonstrated in infected tissues, microscopy could be used to follow experimental inoculations and as an aid in cultivation. And subsequently, the staining procedure proved invaluable as a diagnostic and public health method.

Koch recognized quickly that the peculiar staining properties of the tubercle bacillus indicated something unusual about the cellular properties of the organism. Even in his first work he stated:

It seems likely that the tubercle bacillus is surrounded with a special wall of unusual properties, and that the penetration of a dye through this wall can only occur when alkali, aniline, or a similar substance is present.

The bacteria visualized by my technique show many distinct characteristics. They are rod-shaped and belong therefore to the group of bacilli. They are very thin and are only one-fourth to one-half as long as the diameter of a red blood cell, but can occasionally reach a length as long as the diameter of a red cell. They possess a form and size which is surprisingly like that of the leprosy bacillus.[11]

The tubercle bacillus is constantly present when tuberculosis is present

Using his newly developed staining method (Figure 14.1), Koch proceeded to make a careful microscopic survey of tuberculous tissues. As noted, he had access to excellent clinical material at the Charité Hospital, which was near his institute:

> In all tissues in which the tuberculosis process has recently developed and is progressing most rapidly, these bacilli can be found in large numbers. They ordinarily form small groups of cells which are pressed together and arranged in bundles, and frequently are lying within tissue cells. . . . Many times the bacteria occur in large numbers outside of cells as well. Especially at the edge of large, cheesy masses, the bacilli occur almost exclusively in large numbers free of the tissue cells.
>
> As soon as the peak of the tubercle eruption has passed, the bacilli become rarer, but occur still in small groups or singly at the edge of the tubercle mass, with many lightly stained and almost invisible bacilli, which are probably in the process of dying or are already dead. . . .
>
> Because of the quite regular occurrence of the tubercle bacilli, it must seem surprising that they have never been seen before. This can be explained, however, by the fact that the bacilli are extremely small structures, and are generally in such small numbers that they would elude the most attentive observer without the use of a special staining reaction. . . .
>
> On the basis of my extensive observations, I consider it as proved that in all tuberculous conditions of man and animals there exists a characteristic bacterium which I have designated as the tubercle bacillus, which has specific properties which allow it to be distinguished from all other microorganisms.[12]

The culture of the tubercle bacillus

Koch was well aware that the *presence* of the organism did not, of itself, indicate that it was the *cause* of the disease entity:

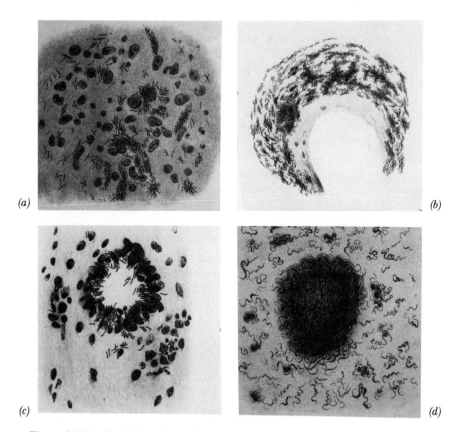

(a)

(b)

(c)

(d)

Figure 14.1   *Koch's drawings of tubercle bacilli in infected tissues. In the original paper, the drawings are reproduced as two-color chromolithographs. The bacteria are blue, the surrounding tissue brown. (a) Tubercle from a section of lung from a patient. (b) Edge of an artery from a case of miliary tuberculosis. (c) Section from a lung showing a giant cell surrounded by tubercle bacilli. (d) Colony of tubercle bacilli as seen in culture.*

In order to prove that tuberculosis is brought about by the tubercle bacillus, and is a definite parasitic disease brought about by the growth and reproduction of these same bacilli, the bacilli must be isolated from the body, and cultured so long in pure culture that they are freed from any diseased production of the animal organism which may still be adhering to the bacilli. After this, the isolated bacilli must bring about the transfer of the disease to other animals, and cause the same disease picture which can be brought about through the inoculation of healthy animals with naturally developing tubercular materials.[12]

This statement, the clearest Koch had yet written of what was to be later considered part of his "postulates" (see Chapter 17), sets out the research protocol that would be needed. But how to culture the organism?

Koch had spent the past six years developing his culture methods for bacterial pathogens, and by this time he had perfected his plate technique (published earlier in the year, see Chapter 11). It must have seemed that it would be a straightforward problem to remove tuberculous tissue from either patients or experimental animals, place the tissue on culture media, and incubate. However, what worked well for anthrax and wound infections did not work for the tubercle bacillus. The organism grew at best very slowly. It did not grow at all at room temperature, while at body temperature the nutrient gelatin that was commonly used liquefied. Also, although the nutritional requirements are not complex, the organism is very sensitive to the presence of inhibitory agents, and therefore behaves as a rather fastidious organism. After many trials, Koch hit on a method using coagulated blood serum. The blood serum served as the source of nutrients, and by coagulating the serum in tubes, it was possible to produce a solid medium upon which colonies could be obtained. By allowing the serum to harden while the test tubes were in a slanted position, a large surface area for culture was obtained (Figure 14.2).

Using this procedure, Koch removed tuberculous tissues from experimental animals and aseptically inoculated serum slants. We can marvel at the persistence he showed in this study. On coagulated blood serum incubated at 37–38°C, no growth of tubercle bacilli was seen at all during the first week, but after the second week, tiny colonies, very small, dry, and scale-like, appeared. If any growth was obtained in the first several days, it was almost certainly due to contaminants and the experiment was considered a failure. Only colonies appearing slowly, at some time during the second week, were likely to represent tubercle bacilli. However, these colonies could generally already be seen at the end of the first week if a 30–40 power microscope was used. After 10–14 days, the culture was transferred to fresh medium using a sterilized platinum wire.

Subsequently, Koch used other nutrient media, and his first use of agar (see Chapter 11) was for the culture of the tubercle bacilli. However, he preferred coagulated serum because the colonies grew more rapidly and showed more characteristic structures.

The pure cultures are virulent

Now came the most critical part of Koch's study—the proof that the pure cultures obtained from tuberculous material caused the disease.

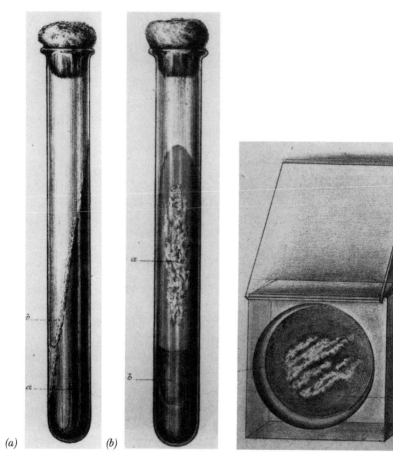

(a)   *(b)*   *(c)*

Figure 14.2   *Drawings from Koch's paper showing tubes and plates for culturing the tubercle bacillus. The original drawings are color lithographs. (a) Pure culture of the tubercle bacillus on coagulated blood serum. (b) The same culture, as seen from the front. The bacilli have not liquefied the coagulum nor grown into it, remaining completely on the surface. (c) A square glass box with a removable cover, containing coagulated blood serum and a culture of tubercle bacilli.*

Here, Koch was greatly benefitted by the fact that guinea pigs can be readily infected with the tubercle bacillus, even though these animals never succumb to tuberculosis naturally:

> A substance can be tested for its virulence by inoculating four to six guinea pigs with it, after making use of all precautions, such as previously disinfecting the site of inoculation, using sterile instruments, etc. The results are uniformly the same. In all animals which are inoculated with fresh

masses containing tubercle bacilli . . . the animals become progressively weaker and die after four to six weeks . . . In the organs of all of these animals . . . the recognizable changes due to tuberculosis occur. That these changes in guinea pigs are due solely to the inoculation of material containing the tubercle bacilli can be seen from experiments in which inoculation was performed with scrofulous glands or fungus masses from joints, in which no tubercle bacilli could be found. In these cases, not a single animal became sick, while animals inoculated with bacilli-containing material always showed an extensive infection with tuberculosis after four weeks.

Up until now my studies have shown that a characteristic bacillus is always associated with tuberculosis, and that these bacilli can be obtained from tuberculous organs and isolated in pure culture. It now remained to prove the most important question, namely, that the isolated bacilli are able to bring about the typical tuberculosis disease process when inoculated again into animals. . . .[12]

When Koch inoculated animals with his pure cultures, they succumbed to tuberculosis, with the same symptoms and pathology seen in animals inoculated with material from human cases.

The results of a number of experiments with bacillus cultures inoculated into a large number of animals, and inoculated in different ways, all have led to the same results. . . .

All of these facts taken together lead to the conclusion that the bacilli which are present in the tuberculous substances not only accompany the tuberculosis process, but are the cause of it. In the bacillus we have, therefore, the actual tubercle virus.[12]

We must consider the work on the tubercle bacillus to be Koch's masterpiece and the culmination of all the work he had done before. It is not surprising that in 1982 during the hundredth anniversary of the publication of this paper, celebrations were held in many countries around the world.

We have seen the evolution of Koch's work clearly through the last few chapters. The speed of this evolution is all the more remarkable when we remember that in 1876, only six years previously, Koch published his first work on anthrax. In those six years he developed a whole series of new techniques which enabled him to discover and characterize the tubercle bacillus.

The properties of the tubercle bacillus make it extremely difficult to work with, and it is remarkable that Koch achieved such quick success in his experiments. However, one who was *committed* to the germ theory of disease would have the *persistence* to continue work when all results were negative. But only one who had gone through the primitive laboratory life of Wollstein would have had the intimate laboratory knowledge necessary to overcome the numerous problems. The tubercle bacillus is extremely tiny, being a tenth the size of the anthrax bacillus. It is very difficult to stain successfully, due to the waxy nature of its cell surface. And it grows very slowly, requiring several weeks for good growth on solid media. Thus, Koch had to work extremely diligently as well as very carefully. If he had thrown out his cultures after one week, he would have been unsuccessful. It was necessary to have *patience and faith!*

Koch was fortunate that the tubercle bacillus is pathogenic not only for humans but for experimental animals. Without an experimental animal which showed characteristic symptoms upon inoculation, his work would have been much harder. He might have cultured the organism successfully, but the actual proof that this organism was the causal agent for tuberculosis would have been much more difficult.

The final link to tuberculosis in the human would, of course, have required human inoculation. But the evidence that Koch adduced was so overwhelming that its acceptance came quickly. Rarely in the field of science has an important work been accomplished so rapidly.

Koch's famous lecture of 24 March 1882[13]

Robert Koch first announced his discovery of the tubercle bacillus at a lecture he gave to the Berlin Physiological Society on the evening of 24 March 1882. It was a memorable occasion. The room where the lecture was given was in the Physiological Institute of the University of Berlin (Figure 14.3). This institute, under the direction of the noted Emil du Bois-Reymond, was a center for scientific excitement.[14]

How did it come about that Koch gave his lecture to a physiological rather than a medical audience? Loeffler[15] has discussed this in some detail. In 1882, German medical research was still dominated by Rudolf Virchow, who was so revered and respected that whatever he said was taken as gospel. (He was once introduced at an important medical meeting as "Professor of Professors".) As we have noted (see Chapter 8), Virchow had been quite cool to Koch's work, and to bacteriological work in general. The establishment of the Imperial Health Office was viewed by Virchow as an affront, and Koch and the other members of

Figure 14.3   *Room where Koch's famous lecture on tuberculosis was given in the Physiological Institute, University of Berlin. This is now the library of the Institute for Medical Microbiology, Humboldt University, German Democratic Republic (East Berlin). Photographed in 1987.*

the Office were shunned by Virchow and his school. Nor was Koch welcomed in the other logical meeting ground for his ideas, the Society for Internal Medicine. But the importance of Koch's work on tuberculosis was so great that it had to be quickly communicated to the scientific and medical world. Was there a suitable forum in Berlin?

Loeffler and Hueppe had become acquainted with du Bois-Reymond and the Physiological Institute because they carried out their animal experiments there. The Imperial Health Office had at the beginning lacked suitable facilities for animal work, and when the laboratory was established, the director, Dr. Struck, arranged for such work to be done at the Physiological Institute. Because of this, Loeffler and Hueppe were made members of the Physiological Society and attended its meetings. According to Loeffler:

> Under the direction of du Bois-Reymond, the Physiological Society pulsed with a vigorous scientific fervor. Everything new was discussed and considered, and above all criticized. The high regard in which new research was held by the Physiological Society meant that any work that was presented there was held in a special light. When Koch discussed with us the best place for him to present his important work, we had no doubt that the Physiological Society would provide the right forum. Koch agreed, and submitted as the title for his lecture "On Tuberculosis". In spite of the vagueness of this title, the meeting room was full to the last place on the night of the lecture. Although not a word of the results of Koch's work had been leaked to the public, all were excited to have the opportunity to hear Koch, the great bacteriologist, in person.
>
> Koch was not optimistic about how his work would be received. On the way to the lecture, he predicted to me that it would take a year of hard battle before his discovery of the tubercle bacillus would be accepted by the medical profession. I, full of optimism and hope, predicted that his work would be quickly accepted.
>
> Koch was by no means a dynamic lecturer who would overwhelm his audience with brilliant words. He spoke slowly and haltingly, but what he said was clear, simple, logically stated—in short, pure unadulterated gold. With increasing excitement, the audience followed every step of his work, examined his excellent preparations, and followed the logic of his experiments. It was quite an evening![16]

Before his lecture, Koch set out on the table his microscopic preparations, more than 200 of them. He had brought everything with him, his microscope, test tubes with cultures, Erlenmeyer flasks, small square glass boxes containing cultures (see Figure 14.2c, these were the days before the Petri plate), pathological material preserved in alcohol. The table was eventually filled with small white plates, each containing a culture. The room was full to overflowing. The lecture proceeded slowly and carefully, Koch laying out for his audience his strongly convincing evidence of the tubercle bacillus.

When Koch finished his lecture, there was silence. No applause, no questions, no debate. Those who attended the lecture remembered later the silence, not of boredom or doubt, but of stunned admiration at the brilliant work they had just heard. Slowly, a few members of the audience rose and shook hands with Koch. Then others began to examine his preparations, his pure cultures, and his other demonstrations. Paul Ehrlich, one of the most important attendees, and an expert on staining bacteria, sank to the stool by the microscope and fell deep into contemplation as he examined the preparations. Later, he was to remark:

> I hold that evening to be the most important experience of my scientific life.[17]

Three weeks later Koch's short paper was published in the Berliner Klinische Wochenschrift[18] (Figure 14.4). It created a sensation throughout the world.

The news of Koch's work spreads to England and the United States

We noted in Chapter 6 that one of Koch's early champions in the English-speaking world was John Tyndall, the eminent British scientist. When Koch's work on tuberculosis was published on 10 April 1882, Tyndall received a copy and on 22 April, only 12 days later, Tyndall arranged to have published an English summary of Koch's work as a letter to the London Times:[19]

> Koch first made himself known by the penetration, skill, and thoroughness of his researches on the contagium of splenic fever [anthrax]. By a process of inoculation and infection he traced this terrible parasite through all its stages of development and through its various modes of action. This masterly investigation caused the young physician to be transferred from a modest country practice, in the neighbourhood of Breslau, to the post of Government Adviser in the Imperial Health Department of Berlin.
>
> From this department has lately issued a most important series of investigations on the etiology of infective disorders. Koch's last inquiry deals with a disease which, in point of mortality, stands at the head of them all. . . . Prior to Koch it had been placed beyond doubt that the disease was *communicable*; and the aim of the Berlin physician has been to determine the precise character of the contagium . . .

Tyndall then proceeded to describe Koch's experiments in detail, taking special pains to describe the extensive animal experiments. And finally, taking a backhanded swipe at the anti-vivisectionists, who were especially strong in England at this time, he noted:

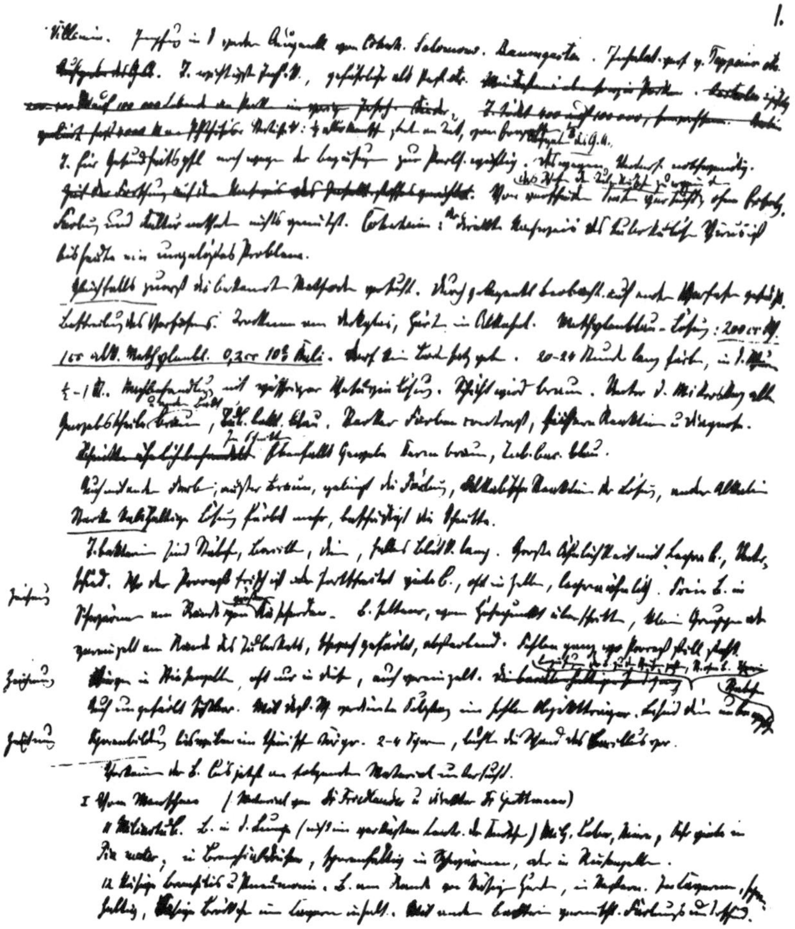

l.

Figure 14.4   *The first page of Koch's notes for the paper he gave on tuberculosis to the Physiological Society 24 March 1882.*

In no other conceivable way [than by animal experimentation] could the true character of the most destructive malady by which humanity is now assailed be determined. And, however noisy the fanaticism of the moment may be, the common sense of Englishmen will not, in the long run, permit it to enact cruelty in the name of tenderness, or to debar us from the light and leading of such investigations as that which is here so imperfectly described.

Hind Head, April 20.

Your obedient servant.
JOHN TYNDALL

In a few weeks, the text of Tyndall's letter crossed the Atlantic and was published in the New York Times.[20] The importance of Koch's work was clearly recognized by the communication media of the day. Publishing Tyndall's letter, the New York Times was disturbed that it had taken so long for the news to cross the waters:

> . . . it is safe to say that the little pamphlet which was left to find its way through the slow mails to the English scientist outweighed in importance and interest for the human race all the press dispatches which have been flashed under the Channel since the date of the delivery of the address—March 24. The rapid growth of the Continental capitals, the movements of princely noodles and fat, vulgar Duchesses, the debates in the Servian Skupschina, and the progress or receding of sundry royal gouts are given to the wings of the lightning; a lumbering mail-coach is swift enough for the news of one of the great scientific discoveries of the age. Similarly, the gifted gentlemen who daily sift out for the American public the pith and kernel of the Old World's news leave Dr. KOCH and his *bacilli* to chance it in the ocean mails, while they challenge the admiration of every gambler and jockey in this Republic by the fullness and accuracy of their cable reports of horse-races.

In the same issue in which it published Tyndall's letter, the London Times had a lengthy editorial, which not only continued Tyndall's discussion about the potential impact of the anti-vivisectionist movement on medical research,[21] but held out the hope that Koch's work would lead to a treatment for tuberculosis.

> If Dr. Koch's investigations and conclusions should be confirmed by further experiments, we shall be able to entertain a reasonable hope that an antidote to consumption and to tuberculous diseases generally may at a not distant date be brought within our reach. It is characteristic of many of the disease-producing bacilli, and probably of all of them, that they can be so altered by cultivation as to produce a mild disease instead of a severe one, and that the designed communication of the former will afford protection against the latter. PASTEUR has lately shown how completely this may be accomplished in the case of the bacillus which causes the splenic fever of cattle [see Chapter 16]; and vaccination itself is now regarded merely as inoculation with the smallpox bacillus, after this has been modifed in its character by being cultivated in the bodies of the bovine race. The experiments of Dr. KOCH, in so far as PROFESSOR TYNDALL describes them, seem as yet to have been carried no further than to the repeated cultivation of the tubercle bacillus in its original virulence; but they will speedily be followed, as a matter of course, by attempts at cultivation in diminished intensity. . . . At this point, therefore, we come into manifest contact with a high probability that the thousands

of human lives which are now sacrificed every year to the diseases produced by bacilli may at no distant period be protected against these for-
midable enemies.[22]

This same idea was quickly picked up by the New York newspapers,
and we note in the New York Tribune for May 3, 1882:

> . . . it is a well-established fact that parasites which produce analogous
> diseases in animals and in the human system can be modifed by cultivation
> until they finally produce a mild form of those diseases, and in this way
> protection may be afforded against virulent conditions. The analogies of
> diseases imply that it may be possible to procure from guinea-pigs or
> rabbits an effective inoculant against consumption, precisely as smallpox
> germs are cultivated in the cow or splenic fever germs in sheep. It is the
> possibility of converting the occupant of a tubercle into a prophylactic
> or preventive agent that lends importance to these interesting experi-
> ments. This possibility is not outined in Professor Tyndall's letter, but it
> may readily be inferred from the vigorous rebuke which he adminsters
> to the anti-vivisectionists in his closing paragraph.[23]

The acceptance of Koch's work on tuberculosis

Koch had predicted (see above) that it would take a year of hard fighting
to convince the medical world of the validity of his work. He was wrong.
Despite a few exceptions, his work was rapidly accepted. The results of
Koch's lecture spread quickly through medical circles of Berlin. Numerous physicians streamed to the Physiological Institute over the next
few days after the lecture to observe the microscopic preparations that
were left in place. Even the eminent Rudolf Virchow appeared, and
although remaining dubious (in his subsequent lectures on tuberculosis
he used the phrase "so-called tubercle bacillus") he did admit that Koch
had something.

As we have noted, tuberculosis manifests itself in the body in a variety
of ways. Pathologists such as Virchow had carefully described, on morphological grounds, these various conditions. Now, examining Koch's
microscopic preparations, Virchow and others could readily convince
themselves that in *all* of these conditions, the same bacterium was present. The discovery of the tubercle bacillus made possible a unified theory
of tuberculosis, in all its manifestations. And because the organism was
so characteristic, it was hard to doubt the validity of Koch's conclusions.

Koch's staining method was quickly improved by Paul Ehrlich and

Franz Ziehl. The same night after the lecture, Ehrlich examined the staining method and showed that the staining properties of methylene blue were strongly improved by addition of aniline (instead of mineral alkali). Further, he discovered the acid-fast character of the staining properties, and on 1 May 1882 reported his results to the Society for Internal Medicine of Berlin.[24] Franz Ziehl (1857–1926), a physician in Heidelberg, discovered that carbolic acid worked better than the aniline used by Ehrlich[25] and Friedrich Neelsen (1854–1894), a physician in Rostock, introduced the use of the dye basic fuchsin (with methylene blue as counterstain).[26] Thus, the Ziehl-Neelsen staining procedure, still widely used, was developed within the first year after Koch's paper. It is also of interest that the Ehrlich's aniline-water solution of gentian violet served as the basis for the development by Hans Christian Gram (1853–1938) of his important bacteriological staining procedure, published in 1884.[27]

Simultaneous development of important discoveries is not uncommon in science. When the time is ripe, many people can make parallel observations. It is thus not surprising that completely independently of Koch, another German physician also discovered the tubercle bacillus. This was Paul von Baumgarten (1848–1928), a Königsberg physician, who observed the tubercle bacillus in unstained preparations of infected tissue. Baumgarten actually reported on his work a few days earlier than Koch, lecturing to the Faculty of Medicine of Königsberg on 18 March 1882,[28] but he did not obtain pure cultures or carry out animal inoculations. Baumgarten's "scoop" can hardly diminish the importance of Koch's contribution.

A number of workers failed to find Koch's tubercle bacillus in material available to them, and some of these workers used their failure to attack Koch and his work. The opposition to Koch was most strong from those who, on theoretical grounds or from morphological observations, had concluded that tuberculosis was not an infectious disease. Koch, always ready to rise in defense of his own work (see Chapter 16), published in 1883 a paper whose main purpose was to silence his critics.[29]

Two of the most outspoken opponents of Koch were H.F. Formad in Philadelphia[30] and A. Spina of Vienna.[31]

Formad not only carried on a continued attack on Koch's work, he went to Berlin to see the staining techniques for himself:

> I found Koch an earnest and conscientious worker, and not as dogmatic and extreme in his views as would appear from his writings; nor is he as

self-satisfied and as rash to jump at conclusions as are some of his followers. Koch has the cooperation of an excellent staff of assistants, all
able mycologists [sic]; but it was a matter of surprise to me that there was
not a single competent pathologist connected with Koch's laboratory; and
such services are evidently much needed, to give to the observations made
there the proper interpretation from a biological and anatomical stand-
point. I was also pleased to learn in Berlin that the discovery of the bacillus
was exaggerated, not so much by Koch himself as by the Imperial Board
of Health, which employs him, and by his over-zealous followers in the
profession. There is strong evidence, however, that Koch's investigations
are biased by the determination to find for each specific disease a specific
fungus. . . . It is really painful to read how some of the younger German
pathologists, and a few of the prominent English surgeons, under the
influence of the bacillus craze, will make, in their publications, assertions
entirely unwarrantable. . . . The only men who attempted to repeat Koch's
experiments, besides [my] work, were Spina and Watson Cheyne. Of the
latter two scientists, Spina came to results entirely different from those
of Koch, and they disprove beyond doubt some parts of Koch's hypothesis. . . . Watson Cheyne, to whom the "British Association for Advancement of Science by Research" had entrusted the investigation of
tuberculosis, and the testing of Koch's researches, did not do justice to
his mission. . . . He went to see some of the different mycologists, con-
sulting only believers in the germ theory . . . He did not inquire, nor did
he care, whether tuberculosis may have any other cause! He simply imi-
tated some of Koch's experiments with the bacillus material in rabbits
and guinea-pigs and obtained, of course, the same results.

Koch in turn analyzed Formad's own work, giving him short shrift:

Formad's self-confidence knows no bounds, as we can see from the following statement he made in the Philadelphia Medical Times for November 18, 1882: "A great deal of good work in pathology is done in America.
Admiration of European pathological works is certainly justifiable, but
there is no reason why the good, honest work of Americans, even that of
young men, should be left unnoticed." In spite of this, Formad failed to
find the tubercle bacillus in the sputum of a large number of patients,
even though these patients had extensive tuberculosis lesions.[32]

Formad's work was also roundly criticized by another American, Edward O. Shakespeare, a Philadelphia pathologist:

There were very many points in Dr. Formad's paper which Dr. Shakespeare believed to be without sufficient foundation. But he would not, at
this time, enter into a general criticism. . . . He intended to limit his remarks to-night to some differences between himself and the author as to
statements made by the latter concerning a recent visit to Koch's labo-

ratory. Dr. Shakespeare also had been in Berlin last summer, and had then enjoyed the privilege for about a month of working under Koch and his assistants during six or seven hours daily. . . . He had not gone to Berlin for the purpose of discovering there the truth or falsity of the claims for the "tubercle bacillus". On the contrary, recognizing the growing importance of research among the various forms of bacteria as possible causes or modifiers of pathological processes . . . he had at length determined to obtain, if possible, ocular demonstration of Koch's classic methods of isolation, culture, and study of minute organisms, and had become one of "the pilgrims" to that Mecca toward which Dr. Formad himself had directed his steps only a few weeks before.

Arrived in Berlin, he had been most cordially welcomed at the *Gesundheitsamt* by Dr. Koch and his corps of accomplished co-laborers, and every possible facility for furthering the object of his visit had been most willingly and courteously tendered during the whole of his stay, though doubtless at the cost of much inconvenience . . . He could say that he had never spent a month with more pleasure or profit. . . .

If Dr. Formad, during the three or four days of his attendance at Koch's laboratory, did not experience an enthusiastic reception, and, as he intimated, was not permitted to experiment upon the pathogenic qualities of the tubercle bacillus, he might far more reasonably have attributed this coldness to an irritation naturally produced by his published remarks in which Koch had been accused of unscientific work, and the insinuation been offered that the researches made at the Imperial Health Office had been unduly influenced by Kaiser Wilhelm, than have assumed from his reception that Koch habitually objected to have any one look into the genuineness and reliability of his work upon tuberculosis. Indeed, the simple fact of his admission at all, under the circumstances, could fairly have been regarded as evidence of Koch's willingness to open his laboratory even to an opponent whom he regarded as unfair. The *Gesundheitsamt* was a department of the German Government. Koch and his chief assistants were officers of the German army or navy.[34] They were all intensely loyal to their Emperor. They believed that Dr. Formad had purposely and unjustifiably stepped outside the proper sphere of a purely scientific communication to publish a reflection insulting to them and their Kaiser. . . .

Dr. Shakespeare took this opportunity to say that his personal observations of Koch, as well as a careful examination of his publications, had led him to the conviction that the whole medical fraternity did not possess a more painstaking, capable, cautious, thoroughly honest and reliable investigator of the causes of disease than the disinguished discoverer of the tubercle bacillus.[33]

But probably the most significant American support for Koch came from E.L. Trudeau, M.D., the most important tuberculosis fighter in the United States:

When in 1882 [Koch] published his epoch-making paper on the Aetiology of Tuberculosis, a paper which stands to-day as a model of its kind, and perhaps the most brilliant publication of a great medical discovery ever written, I was living at Saranac Lake, forty-two miles from a railroad. Nevertheless I learned of his discovery from abstracts in my medical journals. That tuberculosis could be caused by a germ seemed in those days so incredible and startling an assertion that I was most anxious to read Koch's publication; but alas! I could not easily procure a copy from Germany, and if I did I could not read a word of German. I mentioned my dilemma to my friend, Mr. Charles M. Lea, and was overjoyed when six weeks later he handed me a nicely bound English pen and ink translation of Koch's work. . . . The reading of Koch's logical and irrefutable evidence, the clearness of its presentation, and the far-reaching importance of his conclusions made so deep an impression on my mind that, although experimental and laboratory methods were in those days in their infancy and I had never had a lesson of any kind in bacteriology, I decided nevertheless to try to master the necessary technic if possible, and repeat Koch's culture and inoculation experiments. This, after overcoming many difficulties, I at last succeeded in doing. Thus it was Koch's paper that inspired the establishment of the Saranac Laboratory.[35,36]

Spina, on the other hand, attempted to repeat Koch's work, but without success. Koch analyzed Spina's work, showing, among other things, that his microscopy was faulty. Clearly, if an investigator could not *see* the tubercle bacilli, there was no way he could confirm the rest of Koch's observations, and culture attempts would also be doomed to failure. Further, Spina's bacteriological technique was faulty, so that gross contamination occurred. Despite having impure cultures, Spina proceeded to animal inoculations. It was not surprising that such inoculation studies failed.

> To conclude this critical discussion of attempts to repeat my work, we see that the only study that involved a complete analysis of my observations, that of Spina, was quite unsatisfactory. All other papers by opponents of my work produced nothing of value on the etiology of tuberculosis. It has not been a pleasant task for me to criticize this useless literature, but I consider it my duty, in the interest of science, to expose its deficiencies, and hope that in the future more satisfactory work will be published.[37]

Koch need not have bothered to attack his critics! His techniques, improved by Ehrlich and others, soon were being applied widely, and the acceptance of his conclusions by the medical world came quickly.

Even the eminent Virchow eventually came around to the bacterial nature of tuberculosis.

Consolidation of Koch's reputation

The impact of Koch's work on both public health and bacteriology was enormous. By showing that the "Captain of the Men of Death" was a bacterial disease, Koch had made bacteriology one of the most important disciplines of medicine. And by providing a diagnostic procedure of exquisite sensitivity, Koch placed the management of tuberculosis patients on a scientific footing. The hopes raised by the communication media that a tuberculosis vaccine would be soon in coming were, unfortunately, not quickly realized. But, as we shall see in Chapter 18, studies on the immunology of tuberculosis had some important surprises in store, for both Koch and the whole medical community.

Official recognition of Koch's discovery came quickly. On 27 June 1882, only three months after the discovery was announced, Kaiser Wilhelm I appointed Koch as *Kaiserlichen Geheimen Regierungsrat* (approximately translated as Imperial Privy Councilor) (Figure 14.5). With the appointment came an increase in salary, more support for research, and additional assistants.

During the years 1882–1883 Robert Koch worked feverishly at his trade, using his new techniques and his newly found fame to push back the frontiers of medical science. He traveled widely in Europe, spoke at medical congresses, wrote combatively in defense of his work, and provided leadership to his co-workers. He also fought bitterly with Louis Pasteur, as we will discuss in Chapter 16. Especially since his discovery of the tubercle bacillus, visitors from around the world flocked to his laboratory to see his discoveries and learn his new techniques (see Chapter 14).

One important event of 1882–1883 was the *German Exposition of Hygiene and Public Health*. Originally intended to begin in May 1882, the whole exposition was destroyed by fire three days before it was to open, and it had to be rebuilt, opening finally in May 1883. This exposition showed all the latest equipment and techniques of hygiene and public health, and naturally Robert Koch's laboratory was well represented. His exhibit consisted of a whole laboratory equipped for research on infectious diseases and disinfection, showing all the equipment and procedures used at the Imperial Health Office. Actual

Figure 14.5   *Robert Koch, at about the time of the work on tuberculosis.*

demonstrations of pure culture methods and other experimental procedures were carried out continuously during the Exposition, and many visitors, both German and foreign, had the opportunity of familiarizing themselves with Koch's methods. This Exposition certainly had a strong impact on the diffusion of bacteriological knowledge throughout the scientific and medical world and also helped to make Koch's name even more familiar to the general public.

In a letter to his daughter (who was at a boarding school in Thuringia) on 21 May 1883, Koch describes his impressions of the Exposition:

> My dearest Trudy!
> I have been wanting to write you for a long time, especially to thank you for the pretty flowers which you sent me in your last letter, but I have been so busy over the past two weeks attending to the Hygiene Exposition that I simply haven't been able to write. Here in Berlin we also have perfect spring weather. In several days everything will be green. The weather has been simply marvellous for the opening of the Exposition, which has been especially nice since the Crown Prince himself did the honors. The Exposition itself is extremely beautiful and educational. I

wish you were here so that I could take you through the exhibits and explain to you all the pretty things. . . . In our Pavilion we have had a large number of visitors, who have seen our exhibit of dangerous and harmless bacteria and have looked at our photomicrographs and apparatus. I have even had the privilege of explaining bacteria to the Crown Prince, the Grand Duke of Baden, the King and Queen of Saxony, and many other royal personages. . . .

With all my love

Your Papa[38]

The Exposition was not only a scientific and medical success, but also a financial one. During the period from May through October 1883, 900,000 entrance tickets were sold.

Finally, in 1884, Koch completed his final paper on the etiology of tuberculosis, which described in detail his further results since the 1882 work, and which enunciated more precisely the famous "Koch's Postulates"[39] (see Chapter 17). This paper, extensively illustrated with color lithographs of numerous microscopic preparations (see Figure 14.1), was a masterpiece of clarity and logic.

But now, as he had done before, Koch turned to a new and different area for study. His investigations of cholera were to lead him into quite different fields than any previous work and would bring him ever greater fame.

15

The World Traveler: To Egypt and India in Search of Cholera

Up to now, 22 cholera victims and 17 cholera patients have been examined in Calcutta, with the help of both the microscope and gelatin cultures. In all cases the comma bacillus and only the comma bacillus has been found. These results, taken together with those obtained in Egypt, prove that we have found the pathogen responsible for cholera.

—ROBERT KOCH[1]

Robert Koch was about to embark on a new venture. Cholera had broken out in Egypt and threatened to move into Europe again—it had been absent since the Hamburg epidemic of 1866. One of the most dreaded diseases of humans, cholera must not be allowed to enter Europe. Could the new bacteriology help to contain it? The French had established a commission, under the auspices of Louis Pasteur, to go to Egypt. Within days of this news, a German Commission was also established, with Robert Koch as its leader, and on 24 August 1883, Koch and his group reached Alexandria. His work on tuberculosis, begun only two years earlier, was set aside, and he would not return to further work on this disease until 1890 when he began his controversial work on tuberculin (see Chapter 18).

The Background on Cholera[2]

Although cholera had been endemic in India for centuries, it did not spread to the rest of the world until the 19th century. Beginning first

140

in Asia, it spread through Russia and thence to Poland, Germany, Austria, Sweden, and England. A major European epidemic occurred in 1832–1833, and the disease was carried at that time to Canada and the United States by Irish immigrants. From North America, the epidemic spread to Cuba and Mexico. Another pandemic occurred during the years 1846–1862, and it was in this epidemic that John Snow, the British anesthetist, first linked cholera to polluted drinking water. However, Snow's work was ignored. A fourth pandemic occurred in 1864–1875, and it was during this pandemic that Robert Koch first saw cholera as a young physician in Hamburg (see Chapter 3).

The Epidemic of 1883[3]

The mysterious appearance and disappearance of cholera was completely unexplained, but the fear of cholera was widespread and when a new epidemic spread from India to Asia Minor and Egypt in 1883, great concern was expressed in many European countries. The germ theory of disease was now well established, thanks to the efforts of Louis Pasteur and Robert Koch and their collaborators, and it seemed logical that the new bacteriology might be able to contribute something to the control of cholera.

From Damiette, a small port city in Egypt, came the news that cholera had reared its ugly head again and was spreading throughout Egypt. The Egyptian government turned to France and Germany for help. A French team under the auspices of Louis Pasteur was first in the field (see later) and set up their headquarters in the *Hôpital Européen* in Alexandria. The Germans, not to be outdone,[4] quickly put together their own expedition. Under the auspices of the Secretary for the Interior and with extensive support from the Army and Navy, on 9 August 1883 Robert Koch was ordered to lead an expedition to Egypt. Members of the German team included Georg Gaffky, one of Koch's most trusted assistants (see Chapter 10), Bernhard Fischer (1852–1915), another of Koch's assistants, and a Dr. Treskow, chemist in the Imperial Health Office (Figure 15.1). This group departed for Egypt on the evening of 16 August, one week after the expedition had been authorized. Koch was not to return home until nine months later!

The preparations of the German expedition

Bacteriologists can still look in awe at the amazing diligence of the German team. In a week, they assembled a complete travelling bacte-

Figure 15.1   *Members of the German Cholera Expedition of 1883–84. From the left: Gaffky, Treskow (standing), Koch, Fischer.*

riological laboratory, packed it up, and transported it to Egypt. Whereas the French team came prepared primarily for microscopic observations and animal inoculations, Koch's group was prepared for *everything* needed to fulfill Koch's postulates: culture vessels, media, inoculation equipment, sterilization apparatus, etc. A complete inventory of the German expedition's equipment and supplies is given in Table 15.1.[5]

Expenses of the German Commission

The German Commission provided a detailed accounting of the costs which the expedition to Egypt and India entailed. Although the absolute costs have little meaning because of dramatic changes in the value of

Table 15.1 *Catalog of Equipment and Supplies Taken to Egypt by the German Cholera Expedition*

I. Equipment and supplies for microscopy

2 Zeiss microscopes with objectives for low-power and oil immersion
2 hand lenses
3 bottles of cedar oil
600 microscope slides
3000 cover slips
4 dissecting needles
2 fine scalpels
2 fine scissors
2 fine forceps
1 small glass alcohol lamp
20 large watch glasses
20 medium-sized watch glasses
4 small glass funnels
2 adapters for wash bottles
3 tubes Canada balsam
1 leather cloth for polishing
20 camel's hair brushes
100 empty cover slip boxes
1 microtome with freezing attachment and two knives
2 bottles of ether (for the freezing apparatus)
1 sharpening leather
Dyes for staining of microscopic preparations: 50 grams each of methylene blue, methyl violet, gentian violet, ruby, vesuvin, carmin, picrocarmin, picric acid.
Prepared microscope slides of various bacteria, for comparative purposes and for demonstration
500 labels for microscope slides
250 g acetic acid
250 g diluted nitric acid (1:4)
250 g ammonia
250 g potassium hydroxide in solution
250 g glycerol
250 g iodine solution
250 g iodine-potassium iodide solution
250 g aniline
250 g cedar oil
250 g clove oil
250 g anhydrous calcium chloride
2 liters absolute alcohol
250 g ether
150 g glycerol-gelatin
150 g varnish
200 g vaseline

II. Apparatus and supplies for preparation and sterilization of culture media and for culture of microorganisms

1 steam sterilizer
1 water bath

continued

Table 15.1   *Continued*

1 iron tripod
1 three-flame Bunsen burner
1 one-flame Bunsen burner
1 piece fine wire mesh
5 meters gas tubing
3 large brass alcohol lamps
1 small metal alcohol lamp
1 small glass alcohol lamp
1 packet lamp wicks
1 drying oven with thermometer
6 wire baskets
1 iron box to sterilize glass plates
50 glass plates for gelatin cultures
2 crucible tongs
2 wire mesh plates
200 g asbestos
1 sheet of asbestos paper
1 apparatus for solidifying blood serum
100 glass tubes (for solidified serum) with covers
2 cooking flasks
25 Erlenmeyer flasks
50 empty reagent bottles
1 ivory spoon
2 glass funnels
10 glass rods
20 graduated pipettes of various sizes
2 graduate cylinders
2 rolls of cotton
1 piece of gauze
1 piece of parchment (for filtering agar)
2 cloths for pressing out meat extracts
1 packet of glass wool
1/2 ream filter paper
1/2 book litmus paper (red and blue)
1 kg gelatin in tablet form
1/2 kg agar (raw)
500 g peptone
500 g sodium carbonate
1 levelling apparatus
12 graduated pipettes
2 mat-black paper sheets with centimeter-ruled grids (for counting
    bacterial colonies on plates)
10 completely equipped flasks for analysis of air
2 potato knives
8 glass rods with platinum wires
1 box of platinum wire of various sizes
25 sterilized Erlenmeyer flasks with cotton plugs
50 sterilized reagent flasks
80 reagent tubes with sterilized 10 percent nutrient gelatin
40 glass flasks with 50 g nutrient gelatin

continued

Table 15.1 *Continued*

30 reagent tubes with sterilized 2 percent nutrient gelatin containing 1 percent agar
20 glass flasks with the same
80 reagent tubes with sterilized, partly solidified, partly liquid blood serum (cow blood, veal blood, mutton blood)
600 paper labels of various sizes
200 gram gum arabic
20 rubber caps for closing reagent tubes
Dry cultures of chromogenic bacteria for demonstrations
3 maximum thermometers
3 thermometers reading to 360°C
1 thermometer reading to 130°C

III. Instruments, etc., for animal experiments

10 scalpels
12 scissors
12 forceps
4 inoculating needles
4 sterilizable syringes
Sewing needles and thread
1 box with instruments for ocular surgery
10 pieces of heavy wire net (to close mouse jars)
1 mouse forceps
2 dissecting boards for mice
2 dozen fixing needles for the same
4 large glass flasks for collecting preserved organ pieces
47 small flasks for the same purpose, filled with absolute alcohol
Labels from strong paper with string (to label the organ pieces)
2 rolls parchment paper
1 thermometer for measuring animal body temperature

IV. Equipment for various purposes

1/2 kg glass tubes of various sizes
1/2 kg glass rods of various sizes
1 diamond for cutting glass
1 metal file for cutting glass
15 pieces cracking coal
100 corks of various sizes
1 cork borer
20 rubber caps of various sizes
2 meters rubber tubing of various sizes
1 piece gutta percha paper
2 rolls twine
4 pinch clamps
1 piece paraffin
1 meter stick
2 tape measures
1 maximum/ minimum thermometer in case
1 anaeroid barometer
1 tape measure to measure height of ground water

continued

Table 15.1 *Continued*

12 hand towels
12 dust towels
4 bottle brushes
Various tools
Writing materials

V. Drugs and disinfection materials

500 g quinine hydrochloride
250 g salicylic acid
250 g ethyl ether
250 g absolute alcohol
250 g chloroform
250 g potassium chlorate
250 g sodium carbonate
250 g ammonia
250 g ipecac
250 g opium
250 g zinc oxide
250 g castor oil
250 g chloral hydrate solution
150 g bismuth subnitrate
50 g camphor
50 g calomel
50 g menthol oil
50 g sinapol
50 g morphine
1 box with 50 pills of bismuth subnitrate
1 box with 50 pills quinine
1 box with 50 pills morphine
3 boxes with castor oil capsules
500 g carbolic acid (crystallized)
500 g mercury sublimate
250 g 10 percent phenol
2 ivory spoons
10 medicine flasks

VI. Surgical instruments and bandaging materials

2 linen bandages
4 gauze bandages
1 flannel bandage
4 Tripoli bandages
Phenolized silk of various sizes
3 flasks of cat gut of various sizes
1 flask of collodium elasticum
2 pieces of sewing plaster
1 spatula
1 piece English plaster
1 box with sewing and insect needles
1 dissecting apparatus

The above supplies were assembled, packed, and shipped in less than one week's time!

the German Mark over the intervening 100 years, the relative costs are of some interest. The expenses are laid out in the following tabulation:

Equipment purchased for the trip	2265.15
Additional equipment purchased during the trip	896.27
Purchase of reference materials	541.16
Travel costs (train, steamer, etc.)	13,584.67
Hotel costs	10,377.62
Additional travelling expenses	1716.83
Excursions in Egypt and India	1696.39
Payments for personal services	502.29
Purchase of laboratory supplies in Egypt and India	512.50
Miscellaneous laboratory purchases	944.97
Post and telegram	325.66
Exchange rate differences	245.16
Total	33,608.22

At the beginning of the Expedition, the Commission received 6000 Marks for equipment and for travel to Egypt. Additional funds were made available from time to time by the Commission through the German Consulates in Alexandria and Calcutta.

The Egyptian studies

In the short period of a week the whole expedition was fully equipped in Berlin, and Koch and his co-workers left Berlin on the evening of 16 August 1883. Travelling via Munich, Verona, and Bologna, they arrived by train at Brindisi, the Italian port for Egyptian steamers, and departed on the morning of 20 August for Port Said aboard the steamer "Mongolia". The crossing of the Mediterranean, which took four days, was extremely hot. Although the sea was quite calm, Koch got rather seasick. At noon on 23 August the ship anchored at the harbor of Port Said, where the Commission was met by the German Consul. Port Said was under quarantine because of the cholera, and the ship had to wait overnight for clearance. Then the German team was transferred to an Egyptian steamer for the trip from Port Said to Alexandria. Not wasting a moment, even while awaiting clearance, the Commission members occupied themselves with an examination of the state of the cholera epidemic in Port Said. When they reached Alexandria the next day, they found that the French had already set up shop in the Hôpital Européen (see later). Seeking a location where many cholera cases were present, the German team established themselves in the Greek Hospital, where they laid out their laboratory and began their pathological examinations. [There was also a German Hospital in Alexandria, but it was unsuitable since it had very few cholera patients.] Soon after arriving, Koch wrote his wife Emmy:

> Dearest Emmy!
>   Yesterday I arrived here. You have probably already received the postcard I sent you from Brindisi, telling you that because of the quarantine we could not travel directly to Alexandria, but must pass via Port Said, a small city that lies at the entrance to the Suez Canal. We had beautiful weather for the crossing of the Mediterranean, constant sun and a quiet sea. In spite of that, I suffered fearsomely from seasickness. On Thursday afternoon we arrived at Port Said where the Consulate and his assistants received us. We walked for a few hours in the harbor and in the Arab Quarter. In the evening we then went by an Eqyptian steamer to Alexandria. Here also the German Consul met us and arranged for our luggage to be taken to a very good hotel. Unfortunately, the cholera epidemic seems to be waning, but by chance we were able to obtain samples from several cholera patients, as well as from corpses, so that we could begin our studies. I didn't get to bed until midnight last night and this morning I was already up at 5 o'clock. During our few idle moments we have walked in the hotel garden, which is beautiful with palms, blooming oleanders, and giant rubber trees. It is quite hot here, but bearable. This morning we had visits from the Viceroy, the Governor, and the Consul. These formal visits are all rather a pain in this heat. [Presumably, Koch had to put on formal dress clothes for such visits!] For the present I will remain in Alexandria, and you can send me letters in my name direct (Hotel Khedivial, Alexandria, Egypt) or in the name of the German Consulate in Alexandria. So far we are all well and healthy, and hopefully we will stay that way.
> With best wishes,
> Your Robert[6]

German studies on cholera in Egypt

The German team immediately set out to discover the causal organism of cholera. There was no question in their minds that cholera was an infectious disease; all that was necessary was to use Koch's methods to isolate the pathogen. However, despite the success Koch had had with the "difficult" tubercle bacillus, the much simpler "cholera vibrio" eluded him. The work began with microscopical studies of pathological samples from cholera victims but quickly embraced the whole field of medical bacteriology. Culture attempts were made, using a variety of culture media (see the inventory in the table). Extensive animal inoculation studies were carried out, using monkeys, dogs, cats, chickens, and mice, looking for a suitable animal model. Unfortunately, transmission of cholera to animals was not possible (see later), so that these inoculation studies came to naught.

In all cases in which the typical clinical symptoms of cholera were present, a characteristic bacterium was found in the tissue of the intestine (Figure 15.2). However, no sign of a characteristic microorganism was found in the blood, the lungs, the spleen, or the liver. In a letter to the Minister of the Interior on 17 September 1883, Koch reported that he had studied 12 cholera patients at three different

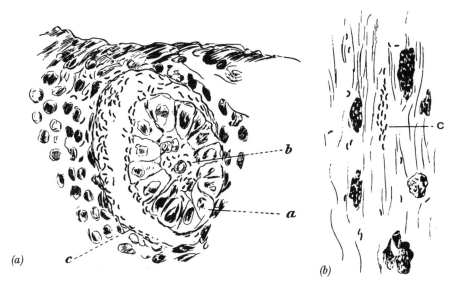

Figure 15.2   *Koch's drawings of microscopic observations. (a) Section of intestinal epithelium from a cholera patient. The comma bacilli are labeled b and c. (b) Cover glass preparation from the intestinal contents of a cholera patient. Characteristic groupings of comma bacilli are labeled in c. (c) Cover glass preparation of cholera stool incubated 2 days on moist linen. The bacteria have multiplied extensively.*

hospitals, had dissected 10 cholera victims, and had found in all cases the same characteristic organism in the intestine. He also reported that he had inoculated 50 mice which he had brought from Berlin, but did not achieve infection. Efforts to culture the organism were also unsuccessful. Koch left open the question as to whether the characteristic bacillus which he saw in the intestine was the causal agent of cholera, although it is clear in retrospect that he had definitely discerned, for the first time, the cholera vibrio. Koch's report was translated into English for the British Medical Journal, which expressed doubt that Koch had found the right organism:

> The whole matter consequently resolves itself into the simple question, whether the bacteria found . . . in the Egyptian specimens of intestine are the bacterial forms which ordinarily accompany disintegrative changes of the tissues, or are they not? . . . we find that Dr. Koch is perfectly sure of the identity of the microorganisms . . . "the same bacilli arranged in the same manner" . . . In the second place, it is clear that, as regards physical characters, the microbes are, practically, indistinguishable from what Koch himself recognises, and continued to look upon for a whole year, as the ordinary accompaniments of putrefaction; and, in the third place, the results of feeding and inoculation experiments on animals failed . . .
>
> It must be clearly borne in mind that Dr. Koch definitely states that he looks upon the experiments hitherto conducted as being merely of an initiatory character; and expressly states that it is not to be assumed that these micro-organisms are the actual causes of the disease, and suggests that further observation may show that they are mere concomitants. For ourselves, judging from the evidence after a very careful perusal of the text, we are inclined to fall in with the latter view, and even to add that, thus far, sufficient data have not been adduced to warrant our assuming that a specific kind of micro-organism has been discovered in cholera. . . .
>
> For our own part, we wish Dr. Koch and his assistants *bon voyage*, and every success in their mission. We do, however, hope that, should the results of their inquiries be of a negative character, they will not hesitate to say so definitely, and with no uncertain sound, as it is a hindrance rather than a help to true progress of medical science that bogey-germs should from time to time be set up, only to be knocked down again, and forgotten.[7]

The day after Koch wrote the report to the Minister of the Interior which has just been discussed, he also wrote a letter to his wife Emmy, describing his experiences in Egypt:

> I am quite well. Cholera has almost disappeared in Alexandria, which for

our purposes is too early. . . . With cholera gone here, I had the idea to travel in upper Egypt, where a fair number of people in the Arabic villages are still dying of cholera, but unfortunately it is dangerous there and therefore not possible. Thus, I am afraid that we are without either cholera patients or cholera victims. If we are to continue our work, we will have to go to India, but I doubt whether the Minister will give us permission to do that. I should be obtaining further instructions from Berlin in two or three weeks and then it will be decided if we are to come home or press on. With luck, I might be home by the middle of next month. I have found a few interesting things, but the main story is still unclear, and seems unfortunately to remain that way. My co-workers are healthy, although we are working very hard, a not small thing in this overbearing heat. As yet we have had not a drop of rain, and often not even any clouds. Simply magnificent blue sky and blazing sun. In the evenings the temperature becomes somewhat more bearable, and in the vicinity of the ocean a fresh, moist wind makes things better. Because of this we have often, of an evening, taken a trip to the beach. We ride on donkeys and we have quite a jolly party when we race across the desert, with the Arab guides stringing on behind. When we finally get to the cliffs, we have a nice fire and a picnic supper. And this is all by the light of the moon! We have also gone out with sail boats into the bay, and now we are planning to spend a few days in Cairo so that we can see the pyramids and the desert. So you see we have plenty to do for amusement.[8]

In addition to his attempts to isolate the causal agent of cholera, Koch carried out epidemiological studies with the goal of describing the conditions that promoted the rise of cholera. He visited quarantine stations at Alexandria, in a branch of the Nile near Damiette, in El Tor, El Wedi, and Suez, in order to obtain some idea of the effectiveness of the quarantine measures. He also examined Moslem pilgrims to determine their role in introducing cholera from Mecca, where the disease was endemic. He also made studies on water supply and filtration, on the influence of the rise and fall of the Nile on the progress of the epidemic, and on correlations with meteorological conditions. Not content to study only cholera, Koch spent his spare time studying amoebic dysentery and infectious conjunctivitis.

Unfortunately, the cholera epidemic in Egypt was subsiding, making it difficult to obtain suitable pathological material. The course of the epidemic for Alexandria (where the data were probably fairly accurate), is shown in Figure 15.3. Over all of Egypt, between 60,000 and 100,000 cholera deaths were reported to have occurred in the 1883 epidemic.

Finally, in the middle of October, Koch received permission from the Minister of Interior to travel throughout Egypt, and later to go to

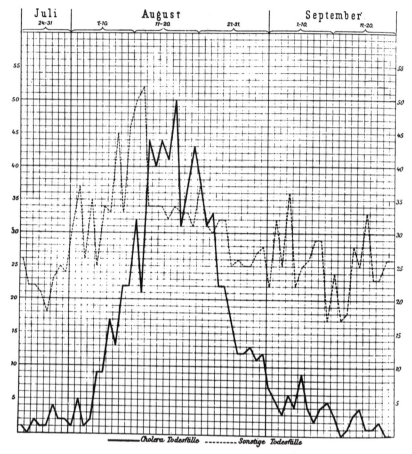

Figure 15.3 *Cholera in Alexandria (solid line). Also shown are total numbers of deaths (dotted line).*

India. From Suez, he wrote to his fifteen-year-old daughter Trudy, who was still in boarding school in Thuringia:

> Dearest Trudy
> Today I am writing you from Suez, the city which lies right at the place where the Suez Canal goes into the Red Sea. Yesterday evening I came here from Cairo by train and this afternoon I am going to El Tor and El Wisch, two tiny harbors on the Red Sea, in order to see pilgrims who have to spend their quarantine here. I won't return to Suez until 8 November, and after that we are going on to Calcutta. In eight days you should get this post card from Suez. So far I have not had any letters

from you. Why aren't you writing? I remain in good health.
        With Love   Your Papa[8]

The French Expedition

As we noted, the French were the first to get a cholera expedition in
the field. The French government (under the auspices of the Department of Trade) appropriated 50,000 francs for support of the expedition on 26 July 1883, and on 15 August the French mission arrived
in Egypt, one day before the Germans departed from Berlin.[9] This ill-
fated group, designated the *Mission Pasteur*, included two assistants of
Pasteur, Émile Roux (1853–1933) and Louis Thuillier (1856–1883), as
well as Isidore Straus (1845–1896), professor of medicine at the University of Paris, and Edmond Nocard (1850–1903), professor of veterinary medicine at Alfort. Occupying the best hospital facilities in Alexandria, the French were equipped primarily for microscopic and animal
inoculation studies.

A total of 24 cadavers were dissected by the French team. They
carried out microscopic examinations of stool specimens and inoculated
experimental animals.[10] Their animal studies were unsuccessful. They
found a variety of different bacteria in their microscopic preparations,
making it impossible to discern the putative pathogen. However, in
blood smears of cholera patients, they saw tiny bodies which they interpreted to be microorganisms and concluded that there was a causal
relation between these bodies and cholera. Later, Koch learned of the
description of these "tiny bodies" and concluded that they were blood
platelets and had nothing to do with cholera.

At about this time, the cholera epidemic subsided and Nocard and
Thuillier turned to a study of rinderpest (cattle plague, now known to
be caused by a filterable virus), another infectious disease which was of
widespread concern. Despite having had no contact with cholera patients for 14 days, Thuillier became ill and died of the disease. It was
a tragic blow to the French team and was especially bitter to Pasteur,
since Thuillier, although only 27 years old, had done yeoman work for
Pasteur during his studies on the anthrax vaccine, including a study in
which Robert Koch was shown to be wrong (see Chapter 16). Before
the French team left for Egypt, Pasteur had written Thuillier:

> Though desiring that you depart for Egypt (because I am confident that
> you know how to protect yourself from all danger, if you will follow exactly

both the letter and spirit of the instructions I have given you), I will be rather sad to see you and Roux leave—you because of the swine fever, Roux because of the rabies.

L. Pasteur[11]

The report of Thuillier's death was reported to *Le Temps* on 21 September 1883 by Émile Roux, and translated into English in *The Lancet*.

> Thuillier and Nocard went on Friday, the 14th, to examine some oxen that had died from the cattle-plague (rinderpest); they returned on Saturday, and on Monday, the 17th, they went to the . . . slaughter yard. . . . Thuillier had a loose stool in the morning; he was, however, lively and in good spirits, and took a bath in the sea, and in the evening we took a drive together. At dinner he ate with a good appetite, and went to bed about half-past ten. He soon fell asleep. About three o'clock in the morning he had another stool, and not feeling very well, he entered his companion's room, saying, "I feel very ill", and then fell prostrate on the floor. Straus and I carried him to his bed; his face was pale and covered with sweat; his hands were cold. . . . We first thought he was suffering from simple indigestion. He soon recovered, took a small draught of opium, and went to sleep. . . . At five o'clock he had a copious watery stool. . . . He vomited his dinner undigested, just as he had taken it the day before. Feeling somewhat relieved, he fell asleep again, after having taken another dose of opium. At seven o'clock he appeared to me to be worse; he complained of feeling cold, and had another motion. Straus and I were obliged to hold him up to prevent his fainting. From this moment everything passed involuntarily; and, in spite of the most energetic treatment, at eight o'clock he was already moribund, with cramps of the muscles of the legs, of the thighs, and of the diaphragm, with alteration of the countenance; nothing, in fact, being wanting to complete the picture of cholera of the most terrible description. We employed strong frictions. All the French and Italian doctors were present. Iced champagne was given freely, and subcutaneous injections of ether were resorted to. In short, everything that could be devised was done to prevent a fatal issue. . . . Notwithstanding all our exertions, he expired on Wednesday morning, the 19th, in a state of asphyxia, which lasted more than twenty-four hours.

Thuillier's body was embalmed and buried the same afternoon in the presence of a large number of mourners. Roux was especially perplexed because none of the team had seen a cholera patient for more than two weeks, and Thuillier himself had taken the greatest care of himself, following Pasteur's precautionary measures to the letter.

Pasteur himself was grief stricken, as Thuillier had been his special favorite.

I felt such esteem and affection for him! I would have gladly asked for him the Legion of Honor! His death was glorious, heroic. Let this be our consolation.

L. Pasteur[12]

Émile Roux corresponded with Pasteur regarding Robert Koch's role at Thuillier's funeral:

M. Koch and his collaborators came as soon as they heard the news. They spoke very beautifully about the memory of our dear departed friend. At the funeral, they brought two wreaths which they themselves fastened to the coffin. "They are only a small token," said M. Koch, "but they are of laurel, and just suitable for one who deserves such glory." M. Koch himself helped carry the coffin.[13]

A number of stories, probably all fantasy,[14] have grown up around the German response to Thuillier's illness and death, all apparently deriving ultimately from Paul deKruif's semifictionalized account in *Microbe Hunters*. One story has Émile Roux hurrying to Robert Koch with the news that Thuillier was dying of cholera. "Would Koch go to see him?" Rumors had already circulated that the French had found the cholera germ and when Koch visited Thuillier, the dying man asked in a weakened voice: "Well, have we found it?" to which Koch, to let him die happy, responded: "Yes, you have found it." According to Howard-Jones, all that is known for fact is that Thuillier did die (on 18 September) and that the German team attended his funeral.

The French team, unsuccessful and disheartened, returned to Paris. The German team went on to India, where Koch proceeded to isolate the cholera vibrio and study it in detail.

On to India

In late October the German Cholera Commission received authorization to travel on to India, and on 13 November the group left Suez. The chemist Treskow returned to Berlin and the Cholera Commission thus consisted of Koch, Gaffky, and Fischer. They originally intended to stay at Bombay, but uncertainty about the availability of study material made Koch decide to go on to Calcutta, the presumed ancestral origin of cholera in the Ganges delta. Travelling on the steamer "Clan Buchanan", the party passed through Colombo (Ceylon) and Madras, arriving at Calcutta on 11 December 1883, Robert Koch's 40th birthday. The trip from Egypt to Calcutta took four weeks. The party was met by the British authorities and excellent laboratory facilities were placed at their disposal at the Medical College Hospital. The laboratory room that they were given was on the second floor of a building adjacent to the hospital (Figure 15.4). The room had three large windows so that it was quite favorable for microscopical studies, but the windows were equipped with blinds which could be closed to keep the sun out, so that the room kept fairly cool. A number of tables, chairs, and cabinets were available, making it possible for the group to set up all of their laboratory equipment. The laboratory even had running water and combustion gas. For studies on experimental animals, a stall in the basement was available, and a chemical laboratory was also placed at their disposal. The autopsy room was nearby, so that fresh material could be quickly brought to the laboratory. The only slight disadvantage was that the Medical College Hospital did not have a large number of cholera patients; however, other hospitals in the area provided all that could be desired in that way.

Gaffky made special note[15] of how fortunate it was that the Commission began its work in the cool time of year, so that the nutrient

Figure 15.4   *The building in Calcutta where Koch carried out his laboratory work on cholera. Photographed in 1983 by Professor Hanspeter Mochmann.*

gelatin plates would remain solid. It was not until the middle of February that the weather became so warm that gelatin cultures could not be prepared.

On the way to India, Koch wrote his wife from the Suez:

> 10 November 1883
>
> We are here in Suez awaiting a ship to take us to India. The ship will apparently come in the morning and then we will be on our way. First we will go to Colombo, on the island of Ceylon, where we will spend several days and try to recover from the two-week sea voyage. And then, on to Calcutta, which will take another 5–6 days. I think that we will arrive in Calcutta in the middle of December.
>
> Today I also wrote to Trudy and sent her stamps, flowers, and a mosquito, the last so that she can learn a little about the dark side of the Orient.[16]

In his letter to Trudy, written the same day, Koch waxed eloquent about the marvellous flower world of Egypt:

> Dear Trudy!
>
> Tomorrow or the next day we will be departing by steamer for India, and I take this time to write you a lengthy letter. I was so pleased to receive your last letter, which Mama sent me. In about two weeks you can write me at the following address: *Geh. Regierungsrath Dr. Koch, Calcutta, German Consulate (British India)*.
>
> I have collected quite a few things for you and send a few of them in this letter. Most important are the stamps, which you wanted so badly. I have obtained some stamps from the post office, but also quite a few which I cut off letters at the German Consulate.
>
> You will also find in this letter a small piece of writing paper with an embossed seal which I obtained from an Arab artist in Damiete. The symbol of this seal signifies your name in Arabic writing, perfectly written out as "Gertrud Koch" . . . The seal itself I will give you when I return.
>
> On this same piece of writing paper I have fastened a strip of paper under which you will find a mosquito, one of the worst kinds that we have here. [Koch's fascination with mosquitoes was to be of scientific use when he studied malaria 15 years later—see Chapter 20.] At night we can only sleep if our beds are completely covered with mosquito netting.
>
> I hope that my letter will not be too heavy if I send you a small souvenir of the tree of the Virgin Mary. This tree, which is really an ancient sycamore tree, stands in Cairo at exactly the place where in ancient Egyptian times the golden city of Heliopolis stood. Now the only thing left of that city are a few stones and a monument, and this tree, of which it is said that the Virgin Mary sought refuge in a cavity on her flight into Egypt. A small fence surrounds the tree, on which is growing oriental jasmin

with snow-white blossoms. I have plucked for you from this tree two leaves and a few blossoms and have dried them. Around here were also pretty plantations of orange trees, the fruit just beginning to ripen, and hedges of flowering myrtle. This should give you a little idea of what winter is like in Egypt! Everywhere there are blooming roses and oleanders. I would have gladly collected more flowers for you, but because of the humid air they are very hard to dry and lose their colors quickly. With this letter I just enclose a small blue flower from the garden of the Hotel du Nil in Cairo. This was the hotel I stayed in. Later I will send you some plants which I found during our camel trip to the desert. Our route will go via Colombo in Ceylon, where we will spend a few days, and then on to Calcutta, which we will reach sometime around the middle of December.

With all my love

Your Papa[17]

On the next day, Koch wrote a lengthy letter to his younger brother Hugo:

Dear Hugo!

I never dreamed, when I left for Egypt, that the expedition would last so long, but I am not especially unhappy about this turn of events. The trip has already brought me enough experiences and impressions to last me for the rest of my life. I have thought of you often, and of my promise to bring you something for your rock collection . . . So I have collected a very pretty limestone that is rich in nummulite fossils [nummulites are foraminiferans] which comes from the Mokokkam Mountains east of Cairo. I found these stones in the small Gizeh Pyramid. The nummulites are about the size of postage stamps. Further, you will be getting a piece from the top of the large Cheops Pyramid. . . . And then I have collected several samples of petrified wood from the large petrified forest in the Arabian Desert east of Cairo, where Europeans don't often travel. . . .

Tomorrow I am leaving by an English steamer for India. Both of my assistants, Staff Officers Dr. Gaffky and Fischer, will accompany me. We travel first to Colombo in Ceylon, stay there two or three days, and then travel on to Calcutta, where we should arrive about 10 December. I would be delighted to hear from you. . . .

With best wishes to you and Doris
Your loving

Brother Robert[18]

On his arrival in Calcutta, he immediately wrote his wife:

Dear Emmy!

I arrived here in Calcutta on the afternoon of my birthday. We had a long hassle with the luggage and making arrangements for a "boarding house". The sea voyage took exactly four weeks, somewhat longer than

I had expected, but it was very pleasant. Our stay in Ceylon was marvelous. Then came a stormy crossing to Madras, where we spent a day, and then five more days to here. The [German] Consul gave me your two letters which had already arrived. I am happy that everything is fine at home. Everything is fine here.

R.[19]

The Research Program in India

During his stay in India, Koch wrote lengthy reports back to the Minister in Berlin, and these reports were printed verbatim in the German newspapers and summarized in English in the British Medical Journal. A short summary of Koch's Indian studies can also be found in Mochmann and Köhler[20] and complete details in Gaffky.[21]

In his first report, Koch set out the research program that he planned to follow:

> I. Microscopical studies of a large number of samples from autopsies, to extend and confirm the Egyptian conclusions regarding the occurrence of a characteristic bacillus in the intestinal tract of cholera patients. Special attention to the specific microscopical properties of this organism to ascertain its characteristic form and size.
> II. Investigation on the occurrence of cholera in animals. Further attempts to inoculate experimental animals of various species with material from cholera patients, using methods of inoculation not yet tried, such as injecting directly into the intestine.
> III. Isolation of pure cultures from the intestine of cholera victims, and use of these pure cultures in animal studies.
> IV. Determination of the biological properties of this bacillus, especially its ability to form spores, its life span, relationship to various culture media, and to a range of temperatures.
> V. Disinfection studies, in order to be able to control the growth of the organism, or to sterilize it.
> VI. Studies on soil, water, and air in relation to presence of choleric materials, especially to answer the question of whether the organism is able to develop outside the human body.
> VII. Special studies on the cholera situation in India . . .[22]

Koch's luck was much better in India than in Egypt. Within days of their arrival in India, Koch and his co-workers had a pure culture of the cholera organism. The secret was the availability of fresh material. The first isolate came from a 22-year old man who had died only 10 hours after the onset of the infection. Less than three hours after death, the body had been autopsied at the Sealdah Hospital in Calcutta, and samples taken for culture. The organism isolated was morphologically identical to that seen by microscopy of intestinal material from cholera patients, both in Egypt and in India. The problem before, of course, had been that the intestine contains such a variety of organisms that, in the absence of a selective culture medium, the likelihood of culturing the pathogen was very low. As we know now, the causal agent of cholera, *Vibrio cholerae*, is not especially fastidious, so that there should be no special problem in obtaining growth in culture, provided other organisms present in the sample do not overgrow the culture. The only real culture requirement is the use of a slightly alkaline pH.

By the time that Koch wrote his report of 2 February 1884, he was quite confident that he had isolated the cholera organism:

> It can now be taken as conclusive that the bacillus found in the intestine of cholera patients is indeed the cholera pathogen. . . . We have determined special properties of the bacillus that make it possible to definitively separate the cholera bacillus from other bacteria. These characteristics are the following: The bacillus is not a straight rod, but rather is a little bent, resembling a comma. The bending can be so great that the little rods almost resemble half-circles. In pure culture these bent rods may even be S-shaped . . . They are very actively motile, a property which can best be seen when examining a drop of a liquid culture attached to a cover slip . . . Another important characteristic is the behavior of the bacteria in nutrient gelatin. Colonies are formed which at first appear compact but gradually spread out as the gelatin is liquefied. In gelatin cultures, colonies of the cholera bacillus can therefore be readily distinguished from colonies of other bacteria, making isolation into pure culture easy. . . . Up to now, 22 cholera victims and 17 cholera patients have been examined in Calcutta, with the help of both the microscope and gelatin cultures. In all cases the comma bacillus and only the comma bacillus has been found. These results, taken together with those obtained in Egypt, prove that we have found the pathogen responsible for cholera.[23]

Illustrations of Koch's results are given in Figure 15.5. Although Koch and his co-workers were confident that they had found the cholera bacillus, others were less sanguine. The British Medical Journal, which carried a running account of Koch's work in India, translated the above report and then went on to say:

> It would be superfluous to say that the statements of such a Commission as this are entitled to the greatest respect; at the same time, it is incumbent on the profession to study the evidence most carefully; indeed, the highest

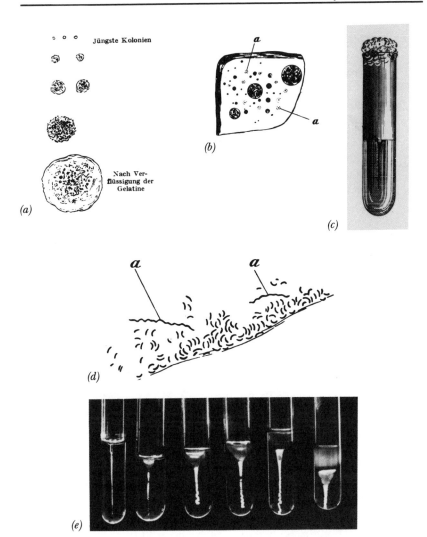

Figure 15.5   *Pure cultures of the cholera vibrio. (a) Koch's drawings of colonies on a nutrient gelatin plate. (b) Lower magnification, showing characteristic colonies (labeled a). (c) Pure culture growing in nutrient gelatin stab. (d) Cover glass preparation of a pure culture, showing the long screw-shaped filaments that sometimes developed (labeled a). (e) Photograph of stab cultures from Gaffky's 1887 paper.*

compliment which can be paid to any Commission of this character is to submit the evidence which it has adduced to early and searching inquiry. . . . It will be remembered that the French Commissioners started with a like idea, and believed that they had found what they were in search of in the blood, whereas the German Commission found it in the intestine. . . . [Indeed], the only evidence adduced by Dr. Koch in support of the unqualified statement that the bacillus which he describes as present in cholera is the veritable cause of the disease, is the circumstance that the micro-organism has not been detected elsewhere. That this particular form of bacterium may be new . . . is . . . by no means improbable; but it is quite possible that the evidence which Dr. Koch adduces in support of its being the *cause* of cholera may receive a wholly different interpretation at the hands of other observers. . . . future observers would do well to bear in mind the remarks made by the *Standard*, in its recent article on Ruskin's "Storm Clouds", "There are not a few researches which might have better stood the test of time had the researchers been less gifted with poetic insight, which gives so large a return of conjecture for so small an investment of fact."[24]

Pasteur was also skeptical that Koch had isolated the cholera pathogen (see Chapter 16). However, this skepticism was mild compared to that of others. Consider the title of the following book, published 10 years after Koch's work:

Robert Koch's comma bacillus is not the cause of cholera. Judgement of an East Indian doctor on the etiology of cholera.[25]

Such skeptics, however, were in the minority, and Koch's work became widely accepted, despite the failure of all animal inoculation studies. In the absence of successful animal studies, Koch turned to epidemiological analysis, which not only gave strong evidence that the comma bacillus was the cause of cholera, but cast important insight into the vehicle by which the pathogen was carried to humans—drinking water. Observations of the connection between cholera and the use of drinking water were made from studies of the so-called Indian "tanks" (Figure 15.6). These tanks served as water storage in many parts of India. They were large square basins which the inhabitants used not only as sources of drinking water, but for bathing and washing clothes. Even the soiled clothes of cholera victims (ripe with fresh bacteria) were washed in these tanks. In one particular tank, 17 cholera deaths were attributed by the Commission to the introduction into the tank of cholera bacteria derived from the clothes of a single cholera patient. The comma bacillus could be cultured from this tank, and the relationship of this tank to the particular cholera epidemic was thus clarified. This study is reminiscent of John Snow's famous study of the cholera epidemic associated with the Broad Street Pump in London. Interestingly, in none of Koch's or Gaffky's writings on cholera is John Snow or his work (published in the 1850s) mentioned, despite the important parallels.

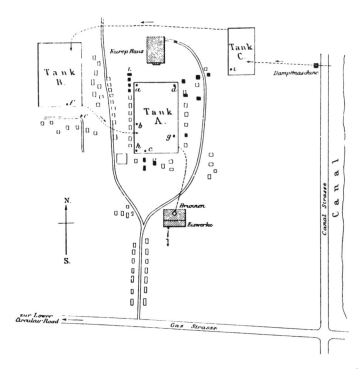

Figure 15.6   *Drawing of a small community that was using drinking water from a concrete tank (Tank A) linked to a cholera outbreak. Koch and co-workers isolated cholera vibrios from this tank.*

The Commission carried out a number of other epidemiological studies in India, all of which are described in detail in Gaffky's 1887 report.[26] Satisfied that they had found what they set out to find, the Commission wrapped up its work at the beginning of March. One reason for stopping work then was that the heat was making it impossible to work with gelatin cultures. The members of the Commission were probably also a little homesick by now. We can get a little idea of the scientific life in India from Koch's letters home to Emmy:

24 December 1883
Everything is fine here. Our work is in full swing and today on Christmas Eve I have been busy with the microscope. We are invited this evening to the German Consulate, although frankly I'd rather spend the evening alone, because I can't get too excited about going in evening dress at Christmas to a strange house. But everyone is so friendly to me here that I can hardly refuse. Hard work I've found a lot of, as lots of people are

dying of cholera, and our work is going very well. I've had a little chance to travel around the city. So far as I can tell, Calcutta is a brutal city. At noon the sun is so downright hot that only an Englishman would go out without a hat. But then in the evenings it gets quite cool, as low as 14–16°C. I have already pulled out my wool socks and in the evening I also wear a light coat. They don't call this winter here. The trees are all leafed out, many are covered with pretty blossoms, and the sun shines so brightly every day that the English and the natives go around with almost no clothes on. . . . However, next month its supposed to start getting warm. We are living in a so-called "Boarding House", a kind of pension. We each have a furnished room and we all eat together at a common table. It's all self-service. We have two Indian "boys", supposedly servants, to wait on us, but they're lazy and good-for-nothing.[27]

Things moved along famously and by the end of January, Koch was feeling rather expansive:

Dear Emmy!

I intended to send off a letter by the last post so that it would reach you in time for your birthday, but I have been so busy that I simply haven't found the time to write. I guess you'll have to treat this letter as a belated birthday wish. . . . Unfortunately, I can't send you the present which I have for you, but I hope that you will welcome it later. I can't tell you what it is. There is a big Exposition here now, which offers the best opportunity to buy very nice things, but I need more money. Although the Exposition has mostly European things, there are quite a few local Indian products and craft items, and when I go to the Exposition I usually visit this Indian Pavilion and fall over myself about all the fantastic things: metal work, jewelry, rugs, shawls, wood work, lots of others. There are also excellent Japanese and Chinese goods. Everything is for sale, and the whole Exposition is nothing more than a large bazaar in which the Europeans and the various Asian peoples show off their prettiest stuff. Imagine a mass of thousands of people, all in their own native costumes, and of every skin color, and then you will have a little idea of what it is like. I have already picked up many nice items, and would buy lots more, but I've run out of money. Would you please send me all the money that you can spare, as fast as possible? The more money, the better, because everything that I bring back you will like. I don't know exactly how much you can send, but a thousand marks wouldn't be too much. I will buy only good things, but they don't seem to be especially cheap. Send the money via the main post office in Berlin to my regular address at the German Consulate.

The last two letters which you and Trudy sent, dated 18 December and 3 January, arrived at the same time. You wrote about Trudy's stay at the boarding school. In the meantime you will have seen from my last letter that I would like it very much if she could come home by Easter.

I presume that she is studying diligently and takes lessons in language, painting, etc. That's fine with me.

We work very hard, but I am also very happy with the results. Unfortunately, it's starting to get hot and in a few weeks it will become hardly bearable. During all of January the weather was like a German summer. On Twelfth Night we were invited by Lord Ripon, the Viceroy, to Barrackpore and there we had a holiday dinner among roses and heliotropes, under a mighty banyan tree. If it gets too hot, we will travel for a few days to Darjeeling, at the foot of the Himalayas. This is the summer resort of Calcutta. It's supposed to be nice. There you can see the eternal snow of the highest mountains on earth.

Concerning my return to Berlin, I can't say anything. For three months now I haven't been able to even think about returning home. We'll have to see what the Ministry decides after it receives my next report. I'll send a letter to Trudy by the next post.

With best wishes

Your Robert[28]

And finally, we have the following letter to Trudy, the last Robert wrote from Calcutta, dated 12 February 1884:

My dearest Trudy!

When you get this letter you will probably already have returned home to your grandparents house. I wrote Mama that I am in agreement that you should leave your boarding school at Easter. However, you must still continue to study and I hope that you will work hard in Berlin on language, music, drawing, and all the other useful things. I would have been quite willing to let you stay another half year at the school, but you must come home to keep Mama company, since I don't know yet when I can come home from India. It doesn't seem likely that I will get back before May, perhaps I must even stay here in India through the summer. However, if I do stay here it won't be in Calcutta, but on the southern slopes of the Himalaya Mountains, where summer weather is bearable. The things you wrote me about life in Calcutta that you read in your geography book are not quite right. During the winter season, from November until March, it is more or less like Germany is during the summer, and then we live just about like they do in Europe. I get up in the morning at seven o'clock, walk or drive to my laboratory, work there until two, come home and eat, then read, write, or visit the Exposition, perhaps take a walk. But lately I haven't been able to take a walk, as the heat has become so fierce. Soon everyone who is able will flee the city for the North, to the Himalayas. We must remain here, living under a punkah. A punkah is a long screen or fan which hangs from the ceiling which the natives keep moving day and night, in order to keep the Europeans cool.

When I am at the Exposition I am very sorry that you aren't here also, to see all the marvelous things that have been brought here from all over

India. I have picked up a few little things for you, however, so that you can get a little idea of what the Indian treasures look like. It's going to be so nice when I unpack my suitcase and bring out one by one the things I've bought you. Then you will see that your Papa really thinks about you all the time. . . . I've also bought some photographs, so that you can see what Egypt and the tropical lands of Asia look like. When we look over these you and I can take a trip together in our imagination, strolling through the ancient forests of Ceylon and through the streets of Calcutta, while I tell you all about the foreign people, their cities and villages, the pretty plants, the amazing animals, the gods of ancient Egypt, and the temples of India. I hope it won't be too long before I see you again and find you as happy and cheerful as you were when I left you. Now, best wishes and a thousand kisses from

Your Papa[29]

Triumphal return to Germany

However, Koch did not have to spend the summer in India, as he had feared. In March permission was received to return to Germany and on 4 April 1884 the Commission left India on the steamship "Bokhara", by way of Bombay, Aden, Suez, and Cairo. They arrived in Egypt on 14 April, ten days after leaving Bombay. They spent a few days in Egypt, partly to learn more about the public health situation there, but mainly because Koch was suffering from malaria and needed a rest from the ocean. The Commission finally left Alexandria for Italy on 22 April. Koch had decided to visit Max von Pettenkofer, the great Munich hygienist, on the way home, in order to present to him the results of the cholera work. Pettenkofer was a believer in a so-called "soil theory of cholera," in which a microbe played only a secondary role. According to Pettenkofer, in addition to the specific organism causing the disease, which he called "x", there was another factor in the soil which contributed to the development of the germ, a factor he called "y" which was essential in order to produce the actual infectious material (called "z") that caused cholera.[30] Based on this hypothesis, Pettenkofer believed that cholera was not transmitted from one person to another, the spread of cholera depended chiefly on the condition of the soil, and drinking water played no role in cholera epidemics. Based on this hypothesis, Pettenkofer taught that the best way of controlling cholera was to remove the local soil factors. Koch's studies, on the other hand, showed clearly that cholera was a contagious disease whose spread could be controlled by public health measures and by avoidance of polluted drinking water. It was understandable then, that Koch would want to present his new evidence on the cholera vibrio to Pettenkofer. The results of Koch's visit with von Pettenkofer, which took place on 29 April, are unreported, but Pettenkofer remained unconvinced.[31]

On 2 May 1884, more than eight months after they left home, the members of the Commission returned to Berlin. Koch, Gaffky, and Fischer received a royal welcome, and were celebrated throughout the whole of the German Empire. Koch, acclaimed as the "Bacillus Father", was received by the Kaiser, had an audience with the Imperial Chancellor Otto von Bismarck, and was awarded a medal.[32] The Kaiser also gave a bronze bust of himself to Koch, and to Gaffky and Fischer he gave his photographic portrait. Koch said years later that the medal was his favorite award, because it had been given to him personally by the Kaiser and because he could wear it like a military decoration. The Commission was given a grant by the Reichstag of 100,000 Marks in gold. On 13 May 1884, the Berlin physicians gave a banquet for the Cholera Commission at the Central Hotel: 700 people attended! Ernst von Bergmann (1836–1907), one of the distinguished professors in the medical school, spoke:

> The *Kreisphysikus* of Wollstein found time, away from his many duties, to carry out important research. In less than a decade, he presented to the world an amazing array of discoveries. Among the many discoveries that he made between the first discovery of anthrax spores and the comma bacillus of cholera, we need mention only one, the discovery of the causal agent of tuberculosis.[33]

Koch did not bring cultures of the cholera bacillus back to Europe. At that time Europe was free of cholera and Koch was reluctant to take the chance of a possible introduction of the disease by means of his cultures. However, in the summer of 1884 cholera broke out in Marseille and Toulon in France, and Koch was sent to investigate. He readily isolated the cholera bacillus from autopsied material and gave demonstrations to two members of the French Cholera Commission, Straus and Roux. The cultures isolated in France were brought back to Berlin and became the basis for cholera research at Koch's institute.

In addition to the preparation of the official report of the German Cholera Commission, which appeared as an *Arbeiten* of the Imperial Health Office,[34] a major conference on cholera was held on 26 and 27 of July 1884. A detailed stenographic record of this conference was taken, which was published in the *Berliner Klinische Wochenschrift*[35] and translated in full into English by the *British Medical Journal*.[36] Koch's triumph at this meeting must have been great, because the eminent Rudolf Virchow, Koch's erstwhile opponent, was present, and at the conclusion of the meeting gave the following statement:

> I say decidedly for myself that, from the beginning, I thought it very probable that the bacillus was, indeed, the *ens morbi*: but, from what I have heard to-day, my conceptions on the subject have arrived at a much greater degree of certainty.[37]

Today, more than 100 years after its etiology was explained by Robert Koch, cholera remains an important public health problem. We will see in Chapter 19, however, how Koch's further work on cholera in Germany, in 1893, played a major role in our understanding of the proper procedures for drinking water purification.

16

The Pasteur/Koch Controversy

When I saw in the program of the Congress that M. Pasteur was to speak today on the attenuation of virus, I attended the meeting eagerly, hoping to learn something new about this very interesting subject. I must confess that I have been disappointed, as there is nothing new in the speech which M. Pasteur has just made. I do not believe it would be useful to respond here to the attacks which M. Pasteur has made on me, for two reasons: first, because the points of disagreement between Pasteur and myself relate only indirectly to the subject of hygiene, and second because I do not speak French well and M. Pasteur does not speak German at all, so that we are unable to engage in a fruitful discussion. I will reserve my response for the pages of the medical journals. (Applause).

—ROBERT KOCH[1]

Throughout most of Koch's career, his relationships with Pasteur were very poor. The reasons for this are complex and difficult to unravel; however, it is clear that the Pasteur-Koch controversy not only influenced the lives of these two outstanding scientists, but also had implications for the development of the microbial sciences. Indeed, even the name for the discipline is linked to the controversy: the Pasteur school preferred "microbiology" whereas the Koch school preferred "bacteriology".[2]

Evidence for the Pasteur-Koch controversy can be found in the scientific writings of these two savants and in their correspondence. The whole controversy has been masterfully outlined by Mollaret in 1983[3] and this paper is summarized briefly here.

As we have discussed (Chapter 10), Pasteur and Koch first met at London in 1881 at the International Congress of Medicine. It was here that Lister arranged for Koch to first demonstrate his techniques and Pasteur exclaimed: "C'est un grand progrés, Monsieur." Pasteur was

59 years of age (Figure 16.1), at the height of his fame and power. He had published his work on fermentations, spontaneous generation, diseases of silk worms, and his studies on wine and beer. He had already carried out some important work on anthrax (which was to be criticized in print a few months later by Koch and Loeffler, see below). He had done his first work on vaccination, and at the London meeting actually spoke on his important discovery of the attenuation of the bacterial agent of fowl cholera and its significance for the whole vaccination process.

Figure 16.1   *Louis Pasteur, photographed at about the time of his conflict with Robert Koch.*

Koch was 38 years old at the time of the London meeting. As we have seen, he did not begin scientific work until relatively late in life, but by 1881 he had published his acclaimed papers on anthrax and on wound infections. He had just developed his very important plate culture technique, which he was about to demonstrate in London. Still in front of him was his most important work on tuberculosis and cholera.

Pasteur did not begin work on anthrax until after Koch's 1876 paper had been published. In fact, Pasteur had already commented favorably on Koch's discovery of spore formation in the anthrax bacillus in a paper he published on 30 April 1877:

> In a remarkable paper, published in 1876, Dr. Koch has shown that the small filamentous bodies discovered by Davaine are able to undergo the formation of refractile bodies . . . which are able to carry the organism over from one year to the next in the absence of growth.[4]

Pasteur's paper presented some of his first work on vaccination with anthrax, work that was to be followed a few months later by the famous experiment at Pouilly-le-fort (see below). At no time did Pasteur make any unfavorable comments about Koch's anthrax work. For the most part, Pasteur ignored Koch's work completely (which may have been partly the reason Koch became so aroused).[5]

Several months after the London meeting, the first volume of the *Mitteilungen aus dem Kaiserlichem Gesundheitsamt* appeared. In addition to the important paper by Koch describing the plate culture technique (see Chapters 11 and 12), there were papers by Koch, Gaffky, and Loeffler attacking Pasteur. Koch not only made a violent attack on Pasteur's discovery of the attenuation of the fowl cholera pathogen, reported at London, but indeed attacked all of Pasteur's work on infection. Koch especially attacked Pasteur's work on attenuation of anthrax, which he had begun as an outgrowth of his work on fowl cholera. Koch accused Pasteur of having impure cultures and of making errors during his inoculation studies.

> Of these conclusions of Pasteur on the etiology of anthrax there is little which is new, and that which is new is erroneous. . . . Up to now, Pasteur's work on anthrax has led to nothing.[6]

One of Pasteur's ideas was that earthworms had something to do with the distribution of anthrax spores throughout the soil, and Koch especially attacked this idea. Most of Koch's attacks on Pasteur were gratuitous, and can only be explained as the young upstart resenting being ignored by the grand master.

Koch's co-workers, Loeffler and Gaffky, were even less polite in their articles published in the same volume of the *Mitteilungen*.[7] This is all the more remarkable when it is realized that Pasteur had already completed his celebrated vaccination experiments for anthrax at Pouilly-le-fort by the time this volume appeared.[8] The basis of the controversy between Koch's group and Pasteur was over the validity of Pasteur's method of attentuation. In order to attenuate his anthrax culture, Pasteur grew the organism at 43°C, a temperature he said prevented the formation of spores. (The implication was that a culture which did not sporulate was attenuated.) However, Koch had found that spores were produced well at 43°C, when cultivated in plate cultures. Since Pasteur did not use plate cultures, the relevance of these observations is unclear. Koch and his co-workers were essentially saying that although Pasteur was successful in his experiment at Pouilly-le-fort, he was successful for the wrong reason and might not be successful another time.

Although Pasteur did not read German, a translation of Koch's paper attacking him appeared in the 20 February 1882 issue of the *Revue d'Hygiène et de Police Sanitaire*, bringing about a violent reaction in French scientific circles. Pasteur responded to Koch's criticism of his vaccination methods by arranging to have some tests run in Germany. These tests, carried out by Pasteur's trusted young assistant Louis Thuillier (who died of cholera the next year; see Chapter 15), have been carefully reported in the correspondence between Thuillier and Pasteur which has been published.[9]

This German test of Pasteur's vaccine came about in the following way. A Dr. Roloff from the Berlin Veterinary School had requested samples of Pasteur's vaccine and Pasteur, wishing to respond to Koch's violent criticisms, proposed the organization of a public experiment in Germany. Roloff obtained permission from the Minister of Agriculture for Prussia, and Thuillier was sent to carry out the experiment. A special commission was appointed to evaluate the experiment, consisting of a Professor Müller from the veterinary school, a Mr. Beyer, the privy counsellor of the Ministry of Agriculture, Rudolf Virchow, and two large landowners, one of whom was also a member of parliament. After his arrival in Berlin, Thuillier wrote Pasteur as follows:

> M. Beyer explained to me that Koch was not a member of the commission, because the Gesundheitsamt, of which he is a member, is an institution

of the German Empire, whereas the experiment is being performed for the Prussian kingdom—and the kingdom and the empire are not to be confused. Judging from the conversations I have heard, Koch's book [the *Mitteilungen*] is not greatly admired.[10,11]

Thuillier arrived in Berlin on 6 April 1882, and the first inoculations were made on an estate in Prussia 150 km from Berlin. This first experiment was essentially a failure, which Thuillier attributed to the use of a different breed of sheep. Pasteur wrote the commission requesting a second experiment, which was begun in early May. This second experiment was successful, and Pasteur's vaccination method became accepted in Germany. Landowners in Germany requesting vaccine for use with their own herds were referred to the commercial source in France.

During his visit to Germany, Thuillier visited Koch and described his visit in detail in a letter to Pasteur. This visit was just after Koch's triumphal report on the isolation of the tubercle bacillus (see Chapter 14). Thuillier reported:

> After reading Koch's paper [on the tubercle bacillus] I could not resist the desire to see these things. M. Koch received me this morning and showed me his preparations, which are quite clear. . . . I wanted to see his animal quarters, which are very clean. I did not find anything to criticize except, perhaps, his method of culture. His repeated cultures do not allow one to be sure that the original seed has been eliminated, since the culture on a solid control spreads out very little and the harvest is taken at the very place where the seed was deposited. But this is not a major criticism.
>
> He has a Zeiss microscope with immersion, and a special lighting system. A movable screen permits him to have an oblique light that shows cellular outlines but not the bacilli. When the screen is removed the light is perpendicular and the stains, in particular the stained bacilli, are very clearly shown. . . . [This was Koch's procedure for distinguishing between the "color" image and the "structure" image; see Chapter 7.]
>
> My visit lasted only an hour and we did not discuss anthrax. M. Koch is not liked by his colleagues. M. Struck is an intriguing ignoramus who has obtained his position as director of the Reichsgesundheitsamt only because he is Bismarck's physician. He is very unpopular and his protégé, M. Koch, shares some of the contempt in which his protector is held. Furthermore, having always lived in a small town of Posen, far away from the scientific centers, [Koch] is a bit of a rustic, and is ignorant of parliamentary language.[12]

In a later letter, Thuillier referred directly to the competition between Koch and Pasteur:

The reply . . . to M. Koch's article [published in the *Mitteilungen*] was received in Berlin last week and was widely circulated. Several people have spoken to me about it. This article makes [our] experiments seem more than ever a direct competition with M. Koch.

Fortunately M. Virchow is a member of the commission and his signature on the report will be testimony before which even M. Koch will have to bow.[13]

Pasteur himself reserved his rebuttal to Koch for the speech he was to give at the IVth International Congress of Hygiene and Demography meeting in Geneva, Switzerland on 5–9 September 1882. At this memorable meeting, Koch was in the audience when Pasteur gave his speech[14]. By this time, Koch was basking in the glory of his discovery of the tubercle bacillus, reported the previous March in Berlin (see Chapter 14). At the conclusion of Pasteur's speech, Koch was asked to respond. The event is reported verbatim in the Proceedings:

Professor R. Koch, of Berlin, took the podium and made the following speech, in German, which was immediately translated into French by M. Haltenhoff:

"When I saw in the program of the Congress that M. Pasteur was to speak today on the attenuation of virus, I attended the meeting eagerly, hoping to learn something new about this very interesting subject. I must confess that I have been disappointed, as there is nothing new in the speech which M. Pasteur has just made. I do not believe it would be useful to respond here to the attacks which M. Pasteur has made on me, for two reasons: first, because the points of disagreement between Pasteur and myself relate only indirectly to the subject of hygiene, and second because I do not speak French well and M. Pasteur does not speak German at all, so that we are unable to engage in a fruitful discussion. I will reserve my response for the pages of the medical journals." (Applause).

M. Pasteur responded to M. Koch that if he had been able to follow the lecture he would have easily understood that new material was presented today. M. Pasteur awaits confidently the reply of M. Koch and will reserve the right to reply to him further at that time.[15]

In a letter to Émile Roux dated 8 September 1882, Pasteur mentions the embarrassing scene at Geneva:

Koch acted ridiculous and made a fool of himself.[16]

And in a letter to his son two weeks later, he again mentions Koch's behavior at the Congress:

It was a triumph for France; that is all I wanted.[17]

According to Mollaret,[18] language problems were at the base of the fight between Pasteur and Koch at the Geneva meeting. Koch could read French (see Chapter 2) but was unable to speak it or to understand it well. During his speech, Pasteur had occasion to refer to some published work of Koch as *Recueil Allemand,* which means collection or compilation of German works. A Professor Lichtheim, a friend of Koch's who was a German-speaking Professor at the University of Berne, was sitting next to Koch to help him with translations. Both he and Koch mistook the word *Recueil* for the word *Orgueil,* which means *pride.* Koch naturally took offense at this presumed attack on "German pride". According to Mollaret, at the words *Recueil Allemand* Koch immediately rose and attempted to interrupt Pasteur, in order to protest this rude attack. Pasteur, not understanding why Koch had jumped up, gestured angrily and silenced Koch. At least part of the basis of the whole controversy was the misunderstanding of language.

The written answer to Pasteur that Koch promised was not long in coming. In a paper with the title: "On inoculation against anthrax. A reply to Pasteur's lecture in Geneva", Koch attacked Pasteur with a violence surpassing even that of his previous paper in the *Mitteilungen.*

> In the program of the 4th International Hygiene Congress in Geneva, which took place in September of this year, Pasteur gave a lecture at the opening session on the attenuation of viruses. As a member of the Congress, I naturally did not want to miss this session, because I was certain that I would learn important new facts concerning Pasteur's technique for attenuation of the anthrax bacillus. . . . But nothing of the sort was to be heard at the Congress. All we had to listen to were new things about fowl cholera, and some details about rabies. Concerning inoculation against anthrax, all that we heard was some completely useless data about how many thousands of animals had been inoculated. . . . All this material served solely as a vehicle for a violent polemic directed against me . . .[19]

The personal viciousness of Koch's attack was shocking. He even went so far as to discredit Pasteur by noting that he was "not even a physician". Koch concluded:

> Although the Congress at Geneva might have celebrated Pasteur as a second Jenner, the members of the Congress should recall that Jenner's triumph did not involve sheep, but human beings.

Pasteur was not long in responding. He answered Koch in an open letter, dated Christmas day 1882, which was preceded by a French translation of Koch's article.[20] In this letter, Pasteur is as personal and polemical as Koch. Of course, more heat than light was revealed in all of these publications.

This Koch-Pasteur controversy, which presumably had as its basis a deep-seated French-German antagonism, carried over into the summer of 1883, when, as we have seen in the previous chapter, competing French and German teams went off in search of the causal agent of cholera. Koch returned from that adventure in triumph, while the French came back empty-handed (and with Thuillier dead). As we noted in Chapter 15, when cholera broke out in France in 1884, Koch went to France to investigate. He readily isolated the cholera vibrio from French patients, the first cultures of *Vibrio cholerae* that were available in Europe. However, Pasteur remained firmly in opposition to Koch, and he strongly protested Koch's visit to France. Thus, in a letter to Straus and Roux in Toulon, Pasteur wrote:

I have been thinking a lot about the big differences between your work and Koch's. He has made very strong but premature conclusions, whereas you have been careful and very reserved. What is the story about Koch's comma bacillus? . . . I haven't seen this organism.

Try to find out the fallacy in his story. How do his microscopic preparations differ from yours? He must have made some sort of great error, if he thinks that in cultures from cholera feces he always sees a bacillus which is never seen in ordinary diarrhea. As much as possible, work by yourself. Keep your cadavers to yourself. The reports which you have received that tell you how great this Koch is are wrong. His knowledge of cholera is not that good. If your results agreed with his, he alone would get all the credit. Already, the German newspapers are crowing![21]

What did Pasteur mean about the German newspapers? The following item, from the Paris newspaper *La Nouvelle Presse*, tells the story.

5 July 1883. *L'affair du Dr. Koch*

We have taken the following from a dispatch from Berlin regarding Dr. Koch's mission to Toulon. "Dr. Koch is going to Toulon at the request of the French government. Because the French Cholera Commission was unsuccessful, the French government has decided to employ exactly Dr. Koch's methods.". . . . It is inconceivable to us that the French government would call into such a mission a Prussian scientist, even one of such scientific authority. France, which has the honor of having in its ranks such eminent savants as M. Pasteur, and which has a Faculty of Medicine renowned throughout Europe . . . does not have any need for the services of a German scientist. M. Pasteur has made a vast number of discoveries of the world of the infinitely small in the past 20 years and is quite able to handle cholera himself.[22]

However, Koch did go ahead and study cholera in France, and apparently did demonstrate his methods to Straus and Roux. Indeed, Pasteur was quite wrong: Koch's work on cholera has stood the test of time (see Chapter 15).

Pasteur's opposition to Koch certainly delayed the use of Koch's plate culture methods in France. On the other hand, when Pasteur first published his results on the rabies vaccine, the Berlin school, under Koch, was opposed to its use. However, public opinion would not permit a rejection of Pasteur's rabies vaccine and a few years later Koch himself organized an antirabies vaccination service, using Pasteur's method exactly. When Koch brought about the establishment in Berlin of an Institute for Infectious Diseases (see Chapter 19), it was conceived in the image of the Pasteur Institute in Paris.

Fortunately, as the years went by, the controversy between Koch and Pasteur gradually subsided, although it never completely disappeared. When Koch announced his discovery of tuberculin in 1890, Pasteur sent Koch a telegram of congratulations (see Chapter 18). However, Koch was absent from the jubilee celebration on Pasteur's 70th birthday, despite the fact that all the rest of the European scientific world was present.

The Pasteur-Koch controversy was certainly rooted, in part, in the French-German antagonism that still festered as an aftermath of the Franco-German War. However, there were also differences in *style* of the Pasteur and Koch schools that, although possibly rooted in national characteristics, have more deep-seated significance. Pasteur, as is well known, was a champion of immunization methods, whereas Koch favored public health measures for the control of disease. Thus, it was Pasteur who introduced vaccination methods, whereas it was Koch who introduced sanitary methods. Pasteur's approach was to treat *individuals* whereas Koch's approach was to treat *populations*. The only time that Koch did venture into the treatment arena, with tuberculin (see Chapter 18), he was seriously in error. One cannot resist commenting that these diverse approaches to the control of infectious disease, by the Pasteur and Koch schools, may also have had their roots in national characteristics of the two countries.

Fortunately, science progresses in spite of personal and nationalistic conflicts!

17

The Berlin Professor

Dr. Koch's new laboratory in the Institute of Hygiene is probably the most complete and perfect of any which exists.

—T. MITCHELL PRUDDEN[1]

Robert Koch's triumph with cholera confirmed to the world that his earlier success with tuberculosis had not been an accident. He soon became one of the most illustrious lights in German science, and his methods and accomplishments attracted visitors from all over the world. He continued to work on cholera, which had not only spread to France but was now also in Germany. This outbreak mobilized the German Army, and Koch organized a series of short courses on cholera for military physicians (Figure 17.1). These courses were also attended by civilian physicians, including those from other countries. In all, physicians from the Austro-Hungarian Empire, Russia, Great Britain, Italy, Spain, Sweden, Luxemburg, the United States, and Australia attended these courses.[2]

These years at the Imperial Health Office were extremely productive scientifically, with Koch's collaborators Loeffler and Gaffky as the most active. During the years 1882 to 1884 Loeffler discovered and characterized the causal agent of diphtheria, *Corynebacterium diphtheriae*. Since at this time diphtheria was one of the major causes of childhood mortality, Loeffler's discovery was of great importance. Indeed, a few years

Figure 17.1 *The class for one of Koch's cholera courses in Berlin, at the Institute for Hygiene. Note the steam sterilizer and the plate-pouring equipment occupying prominent places in the photo. R.J. Petri is seated to the right of Robert Koch who is in the middle of the front row.*

later (see Chapter 18), it led to the discovery by von Behring and Kitasato of diphtheria antitoxin, one of the major discoveries in the field of immunology. Gaffky isolated the causal organism of typhoid fever, another important infectious disease. Gaffky and Koch also made a number of attempts to discover chemical agents that would inhibit the growth of the tubercle bacillus, either in culture or in the experimental animal, but without success. These studies would be the forerunner of the important and controversial studies on tuberculin that are presented in Chapter 18.

The Famous "Postulates"

If Robert Koch is known in bacteriology now, over 100 years later, it is because of the so-called "Koch's postulates" which are attached to his name.[3] Every student in beginning bacteriology memorizes the postulates, and they are frequently written about and discussed in the bacteriological literature.[4] Koch's postulates are essentially a series of steps or procedures that should be followed in order to prove that a specific microorganism is the causal agent of a specific infectious disease.[5] Briefly, the organism must be constantly present in the diseased tissue, the organism must be isolated and grown in pure culture, and the pure culture must be shown to induce the disease when injected into an experimental animal. As emphasized by Carter,[6] Koch used different criteria for establishing causality over the course of his research, and it was not until 1884 that he published the postulates in the form we use today.[7] Indeed, in the early work, Koch emphasized simply the correlation of disease with the *presence* of the organism. Only with the tuberculosis work did he discuss the necessity for pure culture *isolation* of the pathogen.[8] But, it was with the cholera work (see Chapter 15) that the logical structure of an experiment in pathogenic bacteriology was really thrown into relief. As we have seen, cholera is a disease strictly of humans, and Koch thus did not have an animal model in which to reproduce the disease with his pure cultures. Although all the circumstantial evidence convinced Koch that the comma bacillus was indeed the causal agent of cholera, others were not so easily convinced (see later).

Surprisingly, the enunciation of Koch's postulates in their final "textbook" form occurred not in a paper by Koch, but in the paper by Loeffler on diphtheria, dated December 1883. The relevant passage is:

> If diphtheria is a disease caused by a microorganism, it is essential that three postulates be fulfilled. The fulfillment of these postulates is necessary in order to demonstrate strictly the parasitic nature of a disease:
> 1) The organism must be shown to be constantly present in characteristic form and arrangement in the diseased tissue.
> 2) The organism which, from its behavior appears to be responsible for the disease, must be isolated and grown in pure culture.
> 3) The pure culture must be shown to induce the disease experimentally.[9]

Here indeed are Koch's postulates, but not written by Koch! Koch himself apparently first published "his" postulates in the outline form given here in 1890 in his paper announcing the discovery of tuberculin (see Chapter 19). When he was unsuccessful in establishing an animal model for cholera (see Chapter 15), the postulates were thrown back at him by his critics. Thus, an anonymous editorial appeared in the British Medical Journal (Volume II for 1884, page 427) following the English translation of the proceedings of the First Cholera Conference held in Berlin (see Chapter 15):

> Of course, the whole point turns on whether Dr. Koch has made out that the comma-bacillus is really the cause of the disease. In order to demonstrate that a given bacterium is the cause of a disease, it must be proved:

(1) that a special bacterium with definite characteristics marking it out from other forms of bacteria, is constantly present in the parts affected; (2) that this bacterium is present in sufficient numbers to account for the disease; (3) that it is not similarly associated with other diseases; (4) that this bacterium can be cultivated apart from the body, and that its introduction into lower animals is followed by the same effects as the introduction of the infective material itself.

This is apparently the first time the postulates were published in English. The closest Koch himself came to stating his postulates was the several pages of text that appear in his 1884 paper on tuberculosis. The relevant sentences from that paper follow:

> It was first necessary to determine if characteristic elements occurred in the diseased parts of the body, which do not belong to the constituents of the body, and which have not arisen from body constituents. When such foreign structures have been demonstrated, it is further necessary to ascertain if these are organized and if they show any of the characteristics of living organisms, such as motility, growth, reproduction, and spore formation.
>
> The facts obtained in this [microscopical] study may possibly be sufficient proof of the causal relationship, that only the most skeptical can raise the objection that the discovered microorganism is not the cause but only an accompaniment of the disease. However, if this objection has validity then it is necessary to . . . completely separate the parasite from the diseased organism, and from all of the products of the disease . . . and then introduce the isolated parasite into healthy organisms and induce the disease anew with all its characteristic symptoms and properties.[10]

Note that these lines are quoted out of context, and with extraneous material deleted. Koch never made as straight a statement of his "postulates" as that found in Loeffler's paper or in the British Medical Journal.[11]

A key requirement of the postulates which can often not be fulfilled is that for animal inoculation studies. Koch himself was certainly aware of the difficulties this sometimes presented. When he wrote his first paper on cholera following the Berlin conference,[12] he discussed this specifically:

> The comma bacillus is a specific bacterium, found exclusively in association with Asiatic Cholera. As long as this statement is not contradicted, all of my conclusions regarding the diagnostic utility of this bacterium and its relation to the pathology of cholera remain valid. However, my opponents state that a causal relationship between the comma bacillus

and cholera has not been established because it has not been possible to induce cholera artificially in experimental animals. This objection is not valid because Professors Rietsch and Nicati succeeded, during the last cholera epidemic in Marseille, to bring about cholera-like symptoms in dogs and guinea pigs, provided the bile duct had been tied off and a pure culture of the comma bacillus injected directly into the duodenum. . . . These studies have been recently repeated here in my laboratory, using a pure culture diluted so much that less than one-hundredth of a drop was injected. . . . With few exceptions, the treated animals died in 1.5–3 days. . . . These animal studies have been extended in other directions and have shown without a doubt that the comma bacillus is pathogenic. Under these circumstances, it would be very advisable to avoid completely any inoculation studies in humans (as has been recently discussed), and restrict all inoculation studies to guinea pigs and other experimental animals.[13,14]

The difficulty of developing an adequate animal model to fulfill Koch's postulates is one that has plagued medical researchers ever since Koch's first work. In cases where an adequate animal model is not available, one often relies on accidental laboratory infections in humans. The situation with cholera is discussed in some detail by Wilson and Miles,[15] who note a case in which a physician attending the bacteriology course at Koch's Institute accidentally swallowed cholera bacteria and developed the disease. Since there was no cholera at the time in either Berlin or the rest of Germany, it is practically certain that the infection was contracted from the laboratory culture.

It is certainly the case that Koch's postulates have become enhanced through the years by textbook writers and others who read only secondary sources.

The Institute of Hygiene

Although Koch's position at the Imperial Health Office was secure, it lacked status in German medicine. Then as now, the pinnacle of success for a medical researcher was a university professorship. In August 1884, Koch was offered a position as Professor of Pathological Anatomy at the University of Leipzig. Koch's former mentor Julius Cohnheim (see Chapter 6) had just died, and Koch was naturally considered the most suitable replacement. However, Koch turned the position down, stating that he felt he owed it to Prussia and to the German Empire to stay in Berlin.[16]

But then, a short while later, a position became available that Koch did find suitable. A new Institute of Hygiene was to be established at the University of Berlin, and Koch, of course, was offered the position. Arrangements were made so that he could also retain an affiliation with the Imperial Health Office, and on 12 May 1885 Koch was appointed Professor of Hygiene in the Medical Faculty of the Friedrich-Wilhelm University of Berlin. He also retained his title of Privy Councillor.

The background of the founding of the Institute of Hygiene is interesting, since it provides insight into how the whole discipline of public health became established. The first Institute of Hygiene anywhere in the world was founded in 1878 by Max von Pettenkofer at the University of Munich. Pettenkofer practiced what would now be called *environmental medicine*, but with little attention to the significance of bacteriology. As we saw in Chapter 15, Pettenkofer did not believe in Koch's discovery of the causal agent of cholera and actually drank a culture to prove that the bacterium was harmless. (He lived, but Koch was right anyway.)

But it was Koch, rather than Pettenkofer, who brought a modern focus on the discipline of hygiene as a scientific study. After Koch's discoveries, interest in hygiene and public health ran high. It was natural to consider how hygiene should be integrated into the medical curriculum and into medical research. Two schools of thought developed, the school dominated by Koch which believed that hygiene should be a separate discipline, and the school dominated by Rudolf Virchow which believed that hygiene was simply a part of medicine and should not be singled out for special attention in the academy. Virchow's opposition to Koch's new Institute of Hygiene, however, seemed to be more of a "turf" problem than a medical or scientific one, as revealed by his following statement:

> Gentlemen, we in the university are not as confident about hygiene as are those outside the university [that is, Koch and the others at the Imperial Health Office]. As far as we are concerned, hygiene is in the same category as forensic medicine, an applied science which has neither its own methods nor its own concepts. It is of course very important for hygiene to be studied. But we hold it to be even more important that the hygienist study chemistry and microscopic methods in a College of Medicine rather than a College of Hygiene.

Virchow went on to push strongly for the expansion of the present medical institutes, rather than for the establishment of new institutes.

However, his counsel did not prevail. Another Institute of Hygiene was established at the University of Göttingen, in February 1885, under Carl Flügge (1847–1923), a friend of Koch's. And a few months later, the Prussian government established an Institute of Hygiene in Berlin, appropriating funds for the establishment of both a hygiene laboratory and a hygiene museum. It was this Institute, and this laboratory, to which Koch acceded.

The Hygiene Laboratory

Koch's hygiene laboratory was established in an old building formerly occupied by a technical school at Klosterstrasse 32–36 (Figure 17.2). The Prussian government appropriated 60,000 German Marks for reconstruction of the building and the establishment of the laboratory, of which 10,000 Marks was to be for equipment. (To obtain some idea of the value of the German Mark at this time, consider the budget for the German Cholera Commission given in Chapter 15. Since that whole expedition required a little over 30,000 Marks, the 60,000 Marks for

Figure 17.2 *The Hygiene Institute of the University of Berliln, Klosterstrasse 32–36. This was Robert Koch's laboratory from 1885–1891. This building was destroyed during World War II (see Map, Figure 9.1).*

the Hygiene Laboratory does not seem very large.) This laboratory was quickly to become the world center for instruction in bacteriology.

The Hygiene Laboratory was located in the center of Berlin, near the Rathaus, but far from the Medical Faculty (see map, Chapter 9).[17] The building had been built in the 18th century and had then been rebuilt and enlarged in the 1820s. It had a beautiful baroque lecture room on the second floor and an inner court. However, it had been heavily used by the Technical Institute, and by 1876 over 675 students had passed through its doors. In 1884 the Technical Institute moved to new quarters in the suburb of Charlottenburg, leaving the old building empty and available for reassignment by the Prussian government. In addition to the Hygiene Laboratory, the building was to house a Hygiene Museum, containing models, drawings, documents, and other items that had been prepared for the 1883 Berlin Hygiene Exposition (see Chapter 14).

The Hygiene Laboratory opened in this building on 1 July 1885, and the Hygiene Museum on 1 October 1886. Although the building had spacious rooms (Figure 17.3), it had the distinct disadvantage that it was far from the Charité Hospital and medical school buildings. The students had to travel across the city, a thirty minute walk in those days before modern public transportation.

Koch the Professor

Soon after becoming Professor, Koch travelled to Rome to attend a lengthy meeting of the International Sanitation Conference. One purpose of this conference was to develop improved quarantine procedures and to perfect other procedures for the control of infectious disease. Another major issue that was discussed was compulsory inoculation for smallpox. Koch, the passionate traveler, was unhappy that the conference program was so full that he had little time to see the wonders of Italy. He wrote to his seventeen-year-old daughter:

> 6 June 1885
> Rome (Hotel Minerva)
> Dear Trudy!
> Finally I am able to find a little time for letter writing, but only a few moments, so this letter is for both you and your mother. I never suspected that the conference would be so busy. Since the 15th of May I have done nothing from morning to night except work . . . I have had only a little

(a)

(b)

Figure 17.3   *(a) Robert Koch in his laboratory at the Institute of Hygiene. (b) Incubator room in the Hygiene Institute.*

time to see Rome and all of its fantastic art treasures. But now we seem to be nearing the end of our meeting, at least concerning the main tasks, but still no time for relaxation, as now we have to put up with the receptions and other festivities, which are almost as much work as the meetings. Tomorrow morning (Sunday) at 7 o'clock there will be a parade and in the evening the famous *Girandola*, with fireworks. Then on Monday morning at 7 o'clock we are going by boat to Naples. On Tuesday we will go on a navy ship in the Gulf of Naples, Wednesday a trip to Pompei, and Thursday morning back to Rome. . . . As soon as we get back, we will have another round of meetings, which hopefully will be the last. I expect to leave here the end of next week and if all goes well I will be back home in Berlin around the 15–17 June. On the way back I'm going to stop off for a couple of days in Genova. . . . Here in Rome I have been at the German Embassy quite frequently for breakfast with Baron von Keudell. The Embassy is right near the capital, not far from the famous Jupiter temple. The gardens are marvelous, full of palms and orange trees with golden fruit dripping from the branches, reaching out to a sheer rock cliff over which people were thrown in the olden times. Everywhere you look there are remants of the ancient might and glory of Rome. But I dream mostly of returning home where I can sit on our veranda among green trees and drink cool refreshing kefir. . . .

Your loving Papa[18]

Returning to Berlin, Koch set to work right away preparing his lectures and arranging all the short courses. Koch had never given a course of lectures before and he had to spend a lot of time in preparation. He also had to guide the research of his new assistants, as well as to continue to oversee the research of his former assistants at the Imperial Health Office.

In the laboratory on Klosterstrasse, Robert Koch began his professorial duties. He offered three separate courses: a course in Hygiene which met 3 times a week for a select group of students, a laboratory course for the same students, and a series of weekly public lectures on Bacteriological Research Methods, which were given Saturdays at 1–2 o'clock. Koch gave his first public lecture in the Baroque lecture hall on 3 November 1885. In this lecture he gave a historical view of hygiene, and looked forward to the future promise of the discipline.

Koch's lectures and courses were immediately extremely popular, with people coming from all over the world to attend. Although Koch was not a dynamic speaker, he had the reputation of being a careful, thoughtful speaker who spoke clearly and logically. As noted by Martin Kirchner:

Preparation of lectures was a lot of work for Koch. Although as *Kreis-*

physikus and in the Imperial Health Office he had worked in all branches of hygiene, since 1875 his work had dealt primarily with infectious diseases. Nevertheless, he approached his charge vigorously and spent much time preparing drawings, photographs, tables, and models to illustrate his lectures. I attended all his lectures during 1887–89 and can attest that Koch was not the narrow bacteriologist that many claimed him to be. He was a well-rounded hygienist with solid interest in all phases of public health. His lectures were presented with enthusiasm: no one became bored. To enhance his lectures, he took trips with his assistants and students to various hygienic installations, such as water works, slaughter houses, factories, and sewage treatment plants. Because of his reputation and his friendliness, Koch was a popular person. Through his lectures and through example, he strongly stimulated his co-workers and listeners to become excellent hygienists.[19]

We have reports of the status of Koch's laboratory at this time from a number of foreign visitors, of which the following by an American visitor seems especially appropriate to quote here:

> . . . as the greatest advances in recent times have come from the labors of Dr. Koch in Berlin, or from those working under his direction or by his methods; and as his new laboratory in the Hygienic Institute is probably the most complete and perfect of any which exists . . . it will best serve the purposes of the present paper, if the writer gives a general account of the course of study followed here, and the purposes which it is designed to serve.
>
> It is assumed that the student who enters Dr. Koch's laboratory is familiar with the use of the microscope, with the general anatomy of the body, with the minute structure of its different parts as seen under the microscope, with the general methods of preparing diseased tissues for microscopical study, and with the appearance which the different parts of the body present in the various diseases.
>
> The laboratory is a large well-lighted room with tables along the sides for work with the microscope, at which each student has his place before a window, with drawers and cupboards for keeping his apparatus. There are large tables in the centre of the room, supplied with gas and water, for the coarser manipulations, operation on animals, etc. Hoods and hot-air chambers are arranged at the sides of the room and in one corner a small space is partitioned off in which the bacteria which are being cultivated may be kept as much as may be free from dust, either at the ordinary or at an elevated temperature. Adjoining the general laboratory are laboratories for the assistants. Dr. Koch's private laboratory and study are separated from the others by a long hallway. Off from the general laboratory is a room for the janitor and his helpers, in which the apparatus is cleansed and in which is a cremating furnace into which the refuse from the bacterial growths, the bodies of animals which have served the

purposes of experimentation and any infectious material, may be thrown and burned. Large photographic rooms are on the floor above, and also a loft in which smaller animals for experimental purposes are temporarily stored. . . .

Each student is furnished . . . with a set of the necessary apparatus . . . and is given a list of materials which will be used up in his work and which he may procure from the janitor. After cleansing and arranging his apparatus, he is given two or three well known forms of bacteria in the living condition to study. He is to learn all he can about them; their shape and size, whether they are movable or not when living, how they appear when growing, under what conditions they grow best, etc.; and finally with certain forms to see what effect they have when introduced into a living healthy animal. In other words to see whether they are or are not disease-producing or pathogenic. In a word his task is to make out their life history, for himself, as completely as possible. He is aided in this work by Dr. Koch and his corps of well trained assistants. . . .

By repetition of these various manipulations, the student in Dr. Koch's laboratory becomes practically acquainted with a considerable number of the better known forms of bacteria, carrying them through all the different phases of growth, sketching and making notes of his observations. He learns to handle and experiment upon the bacteria causing the most virulent diseases—anthrax, glanders, tuberculosis, etc., without special danger of accidentally communicating them to himself or to others, provided he works intelligently and carefully. He has the opportunity of making analyses of drinking water by which he determines the approximate number of living bacteria in different samples and studies the effects of filtration upon them. He may examine milk and other forms of food which frequently contain living bacteria. He furthermore has the opportunity of carrying on experiments upon the potency of a number of disinfecting agents under varying conditions. . . .

As the end of the course approaches, when the worker has become practically familiar with the methods of research . . . work begins upon the bacterium causing Asiatic Cholera . . .

Finally, the writer cannot refrain from remarking that the calm, judicial mind of Dr. Koch—the master worker in this field—his marvelous skill and patience as an experimenter, his wide range of knowledge and his modest, unassuming presentation of his views are all calculated to inspire confidence in the results of his own work, to stimulate his students to personal exertion in this field, and lend certainty to the already widespread hope that ere long through the resources of science we shall be able to cope successfully with those most terrible and fatal enemies of the human race—the acute infectious diseases.[20]

Another (quite different) view of Koch's institute on Klosterstrasse is the following:

We are in the 3rd story of an old brick and plaster building which would

hardly be deemed worthy of warehouse purposes in America. Herr Koch
has nothing directly to do with our work. He has just now something of
more importance than the teaching of "bugs". Occasionally he passes
through the laboratories and we see his general countenance. The deference with which he is spoken to by all is a surprise to an American. A
German curtsy is something to be studied. . . . I expect it will be something
to *do* before I shall be able to do the proper thing. The bow which is
expected on all occasions begins at the hip joint and is a rigid inclination
of the whole upper part of the body, at a considerable angle. The best
way to practice it is to bind a crowbar from hip to lower spine.

Dr. Pfeiffer is first assistant and really has charge of all the laboratory
work. . . . Dr. Frosch, my overseer, is a genial good natured German. He
is the only one of the assistants who can speak English to any extent.[21]

Marking Time

Now began a period in which Robert Koch consolidated his position as
the leader in bacteriology. No new research programs were begun, but
work was extended on the organisms and diseases which he had already
studied. Loeffler and Gaffky had remained at the Imperial Health Office, but new assistants were hired for the new Institute (Figure 17.4).
In addition to the Bacteriology Laboratory, the Hygiene Institute also
had chemistry and physics laboratories, where research on other aspects
of hygiene were carried out.

Because publication through the Imperial Health Office was no longer
available to Koch in his new position, he started, together with his friend
Carl von Flügge of Göttingen, a new journal entitled *Zeitschrift für Hygiene*.[22] The first issue set out the intentions of the editors for the new
journal.

> Papers on experimental hygiene have in past years been perforce pub-
> lished in only brief form in the weekly medical journals, and it has been
> only possible to publish them in detail in separately published pamphlets.
> These forms of publication have been unsatisfactory both to authors and
> the public. The *Zeitschrift für Hygiene* is intended to fill a real need for a
> publication where detailed experimental protocols, tables of data, and
> other illustrations can be presented so that the field of hygiene can be
> developed to the fullest extent possible.[23]

This journal assured a vehicle for the publication of papers from
Koch's laboratory, as well as those of his co-workers, and soon became
a widely read journal of the *new* bacteriology.

Figure 17.4   *Robert Koch surrounded by his co-workers in the Institute of Hygiene. The picture was probably taken around 1890, just at the time of the discovery of tuberculin and diphtheria antitoxin. The only person not named on the photograph is Proskauer, who is second from the left in the front row. Lower row (from left to right): von Esmarch, Proskauer, Koch, Fraenkel, Pfuhl. Upper row: Pfeiffer, Nocht, Behring, Frosch.*

He was certainly busy during this first year as Herr Professor! By summer of 1886 he wrote Dr. Libbertz, a friend from his youth: "I long for the end of the semester and the chance for a vacation."[24]

He finally did get away for a vacation at the end of July, travelling with Libbertz to Switzerland, and for part of the time with Carl Flügge. His wife and daughter remained home while Koch and his friends did the usual Swiss things, walking in the mountains and climbing on the glaciers. Koch returned to Berlin in early September but immediately set out with his daughter Trudy for Helgoland. As he wrote his wife from Switzerland:

> I want to finish my holiday with some days at the beach and therefore I will go for two weeks to Helgoland. . . . I would like very much to take Trudy with me, and I'll let her decide when to join me.[25]

No mention was made of Emmy Koch joining the pair at Helgoland. By this time, Robert Koch's marriage to Emmy was already in trouble (see Chapter 19).

Through the year 1886–1887 Koch continued his teaching and research at the Hygiene Institute. But no longer was Koch himself immersed in the details of his research. His professorial duties, and the many calls he had from the government and the general public, simply kept him too busy. He took an extended vacation in Switzerland and Helgoland again in the summer of 1887. This time his daughter Trudy, who was 18 years old, went with him to Switzerland as well as to the German sea coast.

Trudy's Engagement

One of Koch's assistants at the Institute was Dr. Eduard Pfuhl (1852–1917), a Captain in the German army medical corps who had been assigned to Koch's staff (Figure 17.5). In October 1887, the engagement of Dr. Pfuhl to Koch's daughter Trudy was announced. Trudy was 19

Figure 17.5   *Eduard Pfuhl, an assistant of Koch who became his son-in-law.*

years old. By this time, Koch was much closer to his daughter than he was to his wife, but he took the engagement well. As he wrote to his friend Carl Flügge:

> As you might well think, I am not completely overjoyed by this news. But the main thing is that for a while at least we will not be separated from our child, and that is a certain consolation. But in several years the young couple will certainly have to leave Berlin, and then the pleasant times will be over.
>
> But unfortunately that is the fate of parents, to be alone in their old age and look for pleasure in the happiness of their children.[26]

Although Koch sounds incredibly old in this letter, he was only 43 and about to embark, although he didn't know it, on an invigorating second marriage! Trudy was married in 1888.

Leading up to tuberculin

About this time, Koch was pleading with his friend Flügge to write a textbook on hygiene. Koch himself was being pressured constantly by publishers for such a book but wished to push this job off to Flügge:

> The longer this task is put off, the more insistent the publishers become. It seems to me very important to fill this important hole in the hygiene literature, if for no other reason than to quiet all the publishers.[27]

At this time, his first and most faithful colleagues, Loeffler and Gaffky, who had stayed at the Imperial Health Office, were leaving Berlin, Loeffler to Greifswald, Gaffky to Giessen. Gaffky returned to Berlin many years later to become the second head of Koch's third institute (see Chapter 20), and Loeffler returned to Berlin to succeed Gaffky. Richard Petri (1852–1921), the curator of Koch's Hygiene Institute (where he invented his famous "plate"), went to the Imperial Health Office to replace Gaffky. Several important new assistants joined the Hygiene Institute including Carl Fraenkel (1861–1915), Richard Pfeiffer (1858–1945), and Martin Kirchner (1854–1925). Fraenkel and Pfeiffer published an important photomicrographic atlas of bacteriology,[28] Pfeiffer did yeoman work on cholera,[29] and Kirchner wrote an early biography of Koch.[30]

By now, Koch's Institute was so full of courses that he had no time to do anything else. But Koch's health was deteriorating and he took a long vacation in Switzerland in the summer of 1889, hoping that it would bring him back to full strength. It is not clear exactly what health problem he had, but it could perhaps have been psychological, associated with the marriage of his daughter and his increasing estrangement from his wife. At any rate, when he returned from his Switzerland vacation in the fall of 1889, he returned to the research laboratory, beginning in a major way the studies on tuberculin that were to put him back into the controversial limelight. We discuss these studies in the next chapter.

18

At the Center of a Storm: Koch's Work on Tuberculin

> *... it is at present much easier to see the bacillus of Koch, than to catch even the most fleeting glimpse of its illustrious discoverer. His name is on every lip, his utterances are the constant subject of conversation, but, like the Veiled Prophet, he still remains unseen to any eyes save those of his own immediate coworkers and assistants. The stranger must content himself by looking up at the long grey walls of the Hygiene Museum in Kloster Strasse, and knowing that somewhere within them the great master mind is working, which is rapidly bringing under subjection those unruly tribes of deadly micro-organisms which are the last creatures in the organic world to submit to the sway of man.*
>
> —A. CONAN DOYLE[1]

For several years after he became Professor of Hygiene, Robert Koch almost ceased doing research with his own hands. The research themes that had occupied the Imperial Health Office were continued, but were carried out by Koch's assistants. He was always available for advice, for discussion, and even for help in the laboratory, but he spent his own time on his lectures and on government commissions. He did manage to find time for socializing over beer and wine, and for extended vacations, but he did no more laboratory work.

But then, sometime in late 1889, Koch's behavior changed. Suddenly he was back in the laboratory again doing work with his own hands. But he was working completely alone, behind closed doors, and for days at a time he would talk to no one. Something important was going on, but the only evidence of his work was the large number of dead guinea pigs which the laboratory helper Meinhardt had to carry out of Koch's laboratory.

And then, in August 1890, the secret of Koch's hidden research was revealed. In a speech for the Tenth International Congress of Medicine held in Berlin, Koch implied that he had found a cure for tuberculosis! The medical world was thunderstruck and the news spread rapidly around the world. The great Koch had triumphed again, but this time in an even more miraculous way. Up to now, bacteriology had given strong insights into the nature of infectious disease but had led to no remedies. In fact, many physicians criticized the "new bacteriology" because it had not led to any successful cures. But now bacteriology had found not only a cure, but a cure for the greatest scourge of humankind.

It is difficult to imagine now how excited the public became about a cure for tuberculosis. As we discussed in Chapter 14, tuberculosis was then the single biggest killer of humans, causing as many as one-seventh of all reported deaths. Tuberculosis in Koch's time was quite different from the disease we have today. In addition to the chronic lung infections we know, many cases of tuberculosis were rampant generalized infections of the body (*miliary tuberculosis*) or ugly, disfiguring skin infections, mainly around the nose and ears (*lupus*). And tuberculosis did not respect social or economic status, affecting rich and poor alike.

And now, the great Koch had found a cure (Figure 18.1). From sanitoria and hospitals all over Europe, thousands flocked to Berlin hoping to get cured. The hospitals could not hold them all, and many stayed in hotels or boarding houses, even on the street, seeking injections of Koch's "lymph".

What was this new treatment that had been so dramatically announced? Koch was not saying. In fact, the only words he said about the treatment in his address to the Berlin Congress were the following:

> [after long study of many chemicals] I have at last found substances which both in the test-tube and in the living body prevent the growth of the tubercle bacilli. All such investigations . . . are very exhausting and slow; and my experiments with these substances, though lasting more than a year, are not yet concluded, so that all I can say at present is that if guinea pigs are treated they cannot be inoculated with tuberculosis, and guinea pigs which already are in the late stages of the disease are completely cured, although the body suffers no ill effects from the treatment. From these experiments I will draw no other conclusion at present than that it is possible to render pathogenic bacteria within the body harmless without ill effect on the body itself.[2]

The tubercle bacillus and delayed-type hypersensitivity

It is one of the interesting accidents of history that the first papers on the two main branches of immunology, *cellular immunology* and *humoral*

Figure 18.1   *Robert Koch as the New St. George, a cartoon appearing at the time of the tuberculin excitement. From "Ulk" of Berlin.*

immunology, were both published in the same year, 1890. The humoral immunology paper was that on diphtheria and tetanus antitoxins, published by von Behring and Kitasato in December 1890.[3] The cellular immunology paper was that presented by Koch at the International Congress.

As we understand the immune response today, antigen stimulation leads to the production of either specific proteins called *antibodies* or specific cells called *activated T cells*. Antibodies, found in the blood and lymph, are specific proteins which interact with antigens in a variety of ways. Activated T cells, derived from the thymus, play a helper role in antibody formation, but also play a direct role in the destruction of cancer cells. Another important role of activated T cells is their involvement in a type of allergic response called *delayed-type hypersensitivity*. It is in this latter process that Koch's tuberculin played a role. Unfortunately, delayed-type hypersensitivity is one of the most complex of immune phenomena, and it has only been in the past several decades that it has been understood by immunologists. Koch's discovery, although of major importance, uncovered a phenomenon so complex that its unraveling was completely impossible by the techniques available in Koch's time.

Many of the symptoms of tuberculosis are a result of the delayed hypersensitivity reaction that is brought about by infection with the tubercle bacillus. What Koch found was that an extract of a pure culture of the tubercle bacillus, when injected under the skin of the guinea pig, brings about a remarkable response if the animal is infected. The active agent is an antigen of the tubercle bacillus which Koch came to call *tuberculin*. After a delay of 24 to 48 hours, the host exhibits either a local or systemic reaction, the severity of which is determined by the amount of material injected and the immunological status of the host. If the host has never encountered the tubercle bacillus, the reaction is mild or almost nonexistent. But in a tuberculous individual, the joint near the site of injection may become swollen, red, and tender, the axillary lymph nodes may enlarge, and fever with a temperature as high as 42°C may occur. The site of the injection may become necrotic and eventually slough off. A similar type of reaction is experienced in the tuberculous human. In severe tuberculin reactions, a patient may experience considerable malaise and may even become prostrate for several days. Pleural or joint effusion may occur.

We know now that the preparation Koch called *tuberculin* contained antigens of the type which elicit delayed-type hypersensitivity in the host.[4]

The background of Koch's tuberculin speech

Most contemporary sources state that Koch did not want to give his speech to the Tenth International Congress, but was pressured to do so by the German government. According to The Lancet:

> Koch, like all scientific men, has his own methods of working, and his own system of declaring his results. He has never yet rushed into print with a discovery until he has been sure of his facts, and all who are in any way acquainted with the circumstances under which *Koch was practically compelled by his Government superiors and by his colleagues to make his premature statement* [italics added] at the International Medical Congress in Berlin will sympathise most deeply with him that he was compelled to break through his usual reticence.[5,6]

In the same issue of The Lancet it is noted that the German Emperor had just conferred on Professor Koch the Grand Cross of the Red Eagle, in recognition of the value of his important discovery. The medal was handed to Koch by the Emperor himself. It was the first time since Alexander von Humboldt that a scientist had received this honor. Koch was also presented the freedom of the city of Berlin, an honor shared by only three other distinguished people, Prince Bismarck, Count von Moltke, and Dr. Schliemann:

> We, the magistrates of this royal city, do hereby declare that
> Privy Councilor of Medicine
> Dr. Robert Koch
> who has raised the knowledge of bacterial life to new heights and demonstrated the true nature of infectious diseases, and using his knowledge to find a remedy in the battle against tuberculosis, has carried the art of healing in new directions, with which his discoveries of natural science exceed those of Jenner, and brought fame and fortune to our city, our suffering citizens offer sincere thanks and hereby bestow on him the freedom of the city.
> Berlin, 21 November 1890

What was Koch's remedy?

Koch's August 1890 paper at the Berlin meeting said absolutely nothing about the source of the remedy, or indeed, anything about its nature or use. (The section dealing with tuberculin was quoted in its entirety above.) The preparation had apparently only been tested on guinea pigs. It is amazing that on the basis of that one paragraph of vague text the whole medical community could rise up in arms, but that they did so attests to Koch's prominence at that time. Much later (see below) Koch revealed that tuberculin was an extract of virulent tubercle bacilli, but for many months after the initial announcement the remedy was kept secret.

The first clinical trials of tuberculin

There was, of course, immediate pressure for Koch to try his new "remedy" on humans, and such trials began almost immediately. Through the fall of 1890, numerous patients were treated at the Charité Hospital and elsewhere. These clinical trials were made primarily under the direction of Dr. Ernest von Bergmann (1836–1907), Surgeon of the Charité Hospital. Actual treatments of patients (not clinical trials, but treatment for a fee) were also undertaken by a student of Koch's, Dr. Georg Cornet (1858–1917). Many so-called "cures" were reported, but because the tests were made under conditions that hardly permitted proper scientific evaluation, the validity of these cures was doubtful. There was no doubt that tubercular patients in general showed strong reactions to injection with tuberculin, and it soon came to be realized that the *Koch phenomenon* (as the hypersensitivity reaction came to be called) had important diagnostic value. But gradually it became clear that tuberculin was useless as a therapeutic agent.

Koch first tested tuberculin on himself, describing his observations in his second paper.[7] He injected into himself 50 milligrams more than the largest quantity that had been given to any patient:

> The symptoms which I myself experienced after an injection of 0.25 cc subcutaneously in my upper arm were as follows: 3 to 4 hours after the injection I experienced pain in the joints, langor, a tendency to cough, and difficulty in breathing, which was followed, in the fifth hour, by a very severe rigor, lasting for an hour, then nausea, vomiting, and a rise of temperature to 39.6°C. This disturbance ceased after about 12 hours, the temperature falling to normal on the next day. The joint pains and lassitude lasted for a few days, and the site of injection remained slightly painful and reddened.

Koch states that initial trials on humans determined that the smallest amount of tuberculin which could elicit a response was about 0.01 cc. He also notes that in healthy and nontubercular individuals, the injection of 0.01 cc hardly had any effect, whereas in the tubercular patient this quantity invariably elicited a marked general as well as a local reaction.

In Koch's second paper cited above, the results of the first human trials are discussed. The rather marked hypersensitivity reactions to the injection of tuberculin were erroneously interpreted as cures. For instance:

> The effect of the injection upon the lupus tissue is to destroy it more or less thoroughly, and cause it to disappear. In some parts the dose may suffice to cause this directly, whereas in others the tissue rather melts or wastes away, requiring repeated injection of the remedy to complete the process.[8]

Koch also noted clearly that tuberculin had no effect on the bacteria, either in culture or in diseased tissue. "It can only affect living tubercular tissue."

Koch's secrecy

Despite the detailed descriptions of the effects on humans, Koch refused to describe the method by which the material was prepared. Indeed, in the second paper he does not even indicate that the remedy is derived from tubercle bacilli. Here is all he says about the nature of the remedy:

> I am not at liberty to describe the origin or preparation of the remedy, as my research is not yet concluded. I reserve this for a future communication. However, doctors wishing to make investigations can obtain the remedy from Dr. A. Libbertz, Lüneburger Strasse 28 II, Berlin, N.W. who has undertaken to prepare it under the direction of Dr. Pfuhl and myself. However, supplies are limited and larger quantities will not be available for some weeks. The remedy is a brown clear fluid which can be stored without special precautions. In use, the liquid is diluted with distilled water. However, the diluted material is liable to bacterial contamination and must be used immediately or sterilized by heat or by addition of 0.5 percent phenol.[9]

The Dr. Libbertz who was preparing tuberculin was an old friend of Koch's from his university days and Pfuhl was his son-in-law. But how available *was* tuberculin from Dr. Libbertz? Conan Doyle visited Libbertz's establishment in November 1890 to inquire about the availability of tuberculin and made the following report:

> I called upon Dr. A. Libbertz, to whom its distribution has been entrusted, and I learned that the present supply is insufficient to meet the demands, even of the Berlin hospitals, and that it will be months before any other applicants can be supplied. A pile of letters upon the floor, four feet across, and as high as a man's knee, gave some indication as to what the future demand would be. These, I was informed, represented a single post.[10]

Robert Koch was severely critized for not revealing the nature of his remedy. Indeed, there was even a German law at that time proscribing so-called "secret medicines" (*Geheimmittel*). According to the British Medical Journal:

The excitement about Koch's new treatment is still at white heat, and has not in the least abated since his publication of last Friday. The nature of the remedy is not known; and though it is supposed to be an organic substance, even this is not certain. Why Koch should have kept his remedy a secret forms the great subject of discussion. . . . We hear on good authority that Koch's demand for a clinic and bacteriological institute met with unexpected opposition, and that he is determined to hold over his secret until all he thinks necessary for the realisation of his scheme shall have been granted him. In the meantime, we are in the presence of the following curious anomalous state of things:—In Germany, where all secret medicines are strictly forbidden by law, and where it is illegal for a medical man to sell drugs, clinical treatment is being actively carried on with a remedy the nature of which is a profound secret, the only person from whom this remedy can be obtained being Dr. Libbertz, a member of the medical profession.[11]

Why did Koch keep the remedy a secret? Koch's decision to keep his remedy a secret was certainly naive, but it was also understandable. At that time, quack doctors applied any remedy to cure any disease, with little control by governmental authorities. What would stop them from attempting to prepare tuberculin? And then, if they were unsuccessful, they would blame Koch rather than the remedy. Koch, attacked repeatedly for not revealing the nature of the remedy, finally explained his stand in an interview to a member of the press:

There is very little use my saying now what the inoculating fluid is or how I have obtained it. It has cost me years of my life, and I propose to retain the secret a few weeks longer from publicity, though it is already known to my assistants and to many of my professional friends. Its preparation demands infinite pains and exactness, and it is being prepared by my assistant, Dr. Libbertz, to whom I have confided this important part of my work, and I believe I am discreet on this subject with good and sufficient reason. Were I to publish now, in the first stage of the discovery, the exact ingredients and the method of preparation of the fluid, thousands of medical men, from Moscow to Buenos Aires, would tomorrow be engaged in concocting it, and injecting it for that matter. Is it far-fetched, then, for me to suppose, as I do, that more than half of these gentlemen are incompetent to prepare the fluid which with special study and special opportunities it has taken me years to prepare? Then these experiments might cause incalculable harm to thousands of innocent patients, and at the same time bring into discredit a system of treatment which, I believe, will prove a boon to mankind.—Then the Professor added earnestly and warmly:—I believe I have the right that the first experiments in its use be made before my own eyes and with the tools which I have made and tested. If these experiments turn out successfully, then the medical world will find me and my elevated assistants only too ready to initiate them into the treatment without the least reserve; but until then, it seems perhaps selfish, but I really claim it as at once our duty and the purest unselfishness, that they must content themselves to be patient.[12]

There is absolutely no evidence that Koch received personal gain, either from tuberculin or from any of his other discoveries. It may be relevant that Louis Pasteur kept secret some of the essential details of his rabies treatment in 1886–1887.[13]

> Medical men are flocking to Berlin from all parts of Europe; at the present about 1,500 have arrived, and it will easily be believed that consumptive patients of all classes clamour for treatment. Dr. Cornet, one of Koch's co-workers, has no fewer than eight temporary consulting rooms in various parts of the city, which are crowded night and day by patients, rich and poor, old and young, from such as have to be carried upstairs to those with only a slight cough.
>
> At a crowded meeting of the Berlin Medical Society yesterday, Virchow defended Koch from the reproach of having made a premature publication. Koch, he stated, at first refused to give the address at the International Medical Congress, only consenting after earnest entreaties had been addressed to him by the Committee and by Dr. von Gossler, Minister of Education.[14]

The minister von Gossler, who was at the center of the controversy, vigorously defended Koch, and took much of the responsibility himself for keeping the nature of the remedy secret. At a meeting of the Prussian Landtag, von Gossler made the following statement:

> On October 27 Koch came to me with the information that he had found a specific against tuberculosis, and asked to be relieved of his duties as director of the Hygienic Institute, in order that he might give all his time to this work. On the day that leave was granted, my conversation with him went deep into this matter. From the first he was ready to tell me all he knew. But I begged him not to do so, as, without any fault of mine, circumstances might arise which might lead to a disclosure on my part, and I might thus do more harm than good. This readiness to disclose everything proves that Koch's motives were purely ideal and scientific.[15]

Koch then volunteered to explain the preparation of the remedy to von Gossler in the presence of two witnesses. But von Gossler felt that it would be hard to make the manner of preparation intelligible to a layperson such as himself. Koch thought that a skilled bacteriologist might be able to learn, in six months, how to prepare the material properly. But he was naturally concerned that the efficacy of all preparations be proved. Since this was not possible by chemical means, the only way the preparations could be known to be efficacious was if they were made exactly by Koch's method. von Gossler then took on himself the responsibility of keeping the manner of preparation secret.

> But to whom did this remedy belong? For my part, there could be no hesitation in deciding that it belonged to its gifted discoverer, Robert Koch. I have always adopted that view with regard to discoveries made by teachers in public institutions—e.g., discoverers of aniline colours, oils, &c. I will, therefore, exercise no official influence over its employment.[16]

von Gossler then went on to explain that neither Koch nor his coworkers were making money on the remedy. It took six weeks to prepare it and a bottle holding 5 grams cost 25 Marks (25 shillings).[17] Because so little was required, the price of a single injection was 0.1 pfennig. However, because of the public outcry, von Gossler agreed that the control of the remedy should be in the hands of the State. An administrative department that would issue the remedy was envisioned. This was presumably modeled after the Pasteur Institute, which had been established just a few years before (1886) for the purpose of dispensing antirabies vaccine.

We can well sympathize with the predicament in which Koch found himself. Almost certainly he had not anticipated the blind enthusiasm with which his announcement was received. The pressures on Koch must have been tremendous, as indicated by the following quotation:

> Whatever may be the ultimate decision as to the system, there can be but one opinion as to the man himself. With the noble modesty which is his characteristic, he has retired from every public demonstration; and with the candour of a true man of science his utterances are mostly directed to pointing out of the weak points and flaws in his own system. If anyone is deceived upon the point it is assuredly not the fault of the discoverer. *Associates say that he has aged years in the last six months, and that his lined face and dry yellow skin are the direct results of the germ-laden atmosphere in which he has so fearlessly lived. It may well be that the eyes of posterity . . . may fix their gaze upon the silent worker in the Kloster Strasse, as being the noblest German of them all.* [italics added][18]

In reaction to the enormous pressures, Koch requested to be relieved of his administrative duties at the Hygienic Institute. The Minister of Education, von Gossler, then arranged, on October 24, 1890, that Koch could step down from his official work at the Hygiene Institute, with Professor von Esmarch being temporarily appointed in his place. However, Koch also requested the establishment of a central department to administer the tuberculin treatment. He wanted a new institute to be established for him, to consist of two parts. One part would be a clinical department with a hospital, where the tuberculosis treatment could be administered. The other would be a scientific institution which would be specially concerned with research on the new remedy. Both institutions would be connected with the Charité Hospital. Koch would be head of the whole, with two assistant directors. No teaching would be done, only research. The Berlin city government also agreed to contribute, as did a private donor.[19]

And this, then, was the origin of Koch's second institute, the Institut für Infektionskrankheiten.

Foreign reception of Koch's discovery

When the news of Koch's discovery of tuberculin reached Paris, Pasteur sent a congratulatory telegram to Koch:

> M. Pasteur and the heads of the departments of the Pasteur Institute send to Robert Koch all their best congratulations for his great discovery.[20]

During a meeting of the French Academy of Sciences, extensive discussions of Koch's discovery ensued. Naturally, Pasteur was at the center of these discussions, and was asked to comment. "Cela y est, cela y est, il n'y a pas á discuter." he responded. The Paris correspondent of the Deutsche Medizinische Wochenschrift, who reported this meeting, added:

> All the world rejoices in the humanitarian significance of Koch's discovery.[21]

The French were even willing to bury their antipathy to the Germans in light of this amazing discovery!

Among the thousands of physicians who flocked to Berlin to see Koch's treatment was Sir Joseph Lister. According to one report, Lister's visit was complicated by the crowds:

> Berlin overrun during winter with hundreds of t.b. patients who rushed

to Berlin for the Koch treatment. Lister of England (then Sir Joseph) brought his niece, was there a week before Koch could see him.[22]

Lister returned to London with enthusiastic reports of the success of Koch's treatment. According to *The Lancet*, Lister "compared the action of Koch's fluid with that used by Pasteur in the case of anthrax, an injection of which gave complete immunity from this disease . . ."[23]

Indeed, Lister's uncritical acceptance of Koch's treatment is amazing, considering the short time that he spent there and the great difficulty of assessing the status of a tuberculosis case. In his own address, given at King's College Hospital, Lister waxed eloquent:

> The effects of this treatment upon tubercular disease are simply astounding. . . . I saw in Berlin a patient with extensive lupus of the cheek, in whom an injection . . . had been performed two days previously. The cheek was enormously swollen and red, and the affected part of the skin covered with crusts of dried serum which had exuded as the result of the intense inflammation that had taken place. Meanwhile, no inflammation had been produced in any other part of the body. . . . Tubercular tissue only is affected by the treatment. . . . But, it may be asked, how far are these effects likely to be permanent, and what limits are to be anticipated to the curative agency of the method? In seeking for an answer to these questions let us return to the case of lupus of the face. It is found that some parts of the diseased tissue lose their vitality altogether under the violence of the local action and are discharged as sloughs. There are other situations where necrotic portions may be so disposed of. . . . While the tubercular tissue is got rid of by this treatment, it appears to be clearly established that the tubercle bacilli are not killed by it. It does not, however, at all follow that the tubercular disease may not be cured, although the bacilli are not directly affected. . . . It is the living tubercular tissue on which Koch's remedy directly operates; yet the bacilli are affected indirectly.
>
> Through Dr. Koch's great kindness I had the opportunity of penetrating into the arcana of the Hygienic Institute of Berlin and seeing most beautiful researches being carried on in that institution, of which Koch is the inspiring genius. I saw things which, while they excited my admiration, made me also feel ashamed that we in this country . . . are so greatly behind our German brethren. The researches to which I . . . refer are still in progress, and fresh facts are accumulating day by day. As they have not yet been published, I am not at liberty to mention details, but there can be no harm in my saying this much, that I saw, in the case of two of the most virulent infective diseases to which man is liable,[24] the course of the otherwise deadly disease cut short in the animals on which the experiments were performed by the injection of a small quantity of . . . an inorganic chemical substance as easily obtained as any article in

the materia medica. And not only so, but by means of the same substance these animals were rendered incapable of taking the disease under the most potent inoculations; perfect immunity was conferred upon them. I suspect that before many weeks have passed the world will be startled by the disclosure of these facts. If they can be applied to man . . . the world will be astonished. . . . At the present time Koch is engaged in the earnest endeavour to produce his remedy for tubercle by some process which could be divulged without the risk of the public being supplied either with material useless from its inertness, or, on the other hand, with deadly poisons. . . . But by publishing now the precise mode of preparing this material, he might do immense harm instead of good . . . And I must say that the carping against Koch on account of what is spoken of as a "secret remedy" can only proceed from absolute ignorance of the beautiful character of the man.[25]

One can be duly impressed with the supreme ignorance in which medical research operated in those heady days!

Among the many noteworthy people who were captivated by Koch's tuberculin was Edward L. Trudeau (1848–1915), the American physician who had established an important tuberculosis sanitorium at Saranac Lake, N.Y. Years later, Trudeau remembered his reaction to tuberculin:

It would be hard to exaggerate the intense excitement that pervaded the little colony of invalids at Saranac Lake when Koch's first announcement of his specific was published in the daily press, and I had all I could do to prevent several of my patients from rushing over to Berlin at once to be cured. The first tuberculin I received came in a small glass bulb, and was sent me by Dr. Osler who, with his usual generosity, shared with me the first bottle of the priceless fluid he had just received from Germany. This small bulb, which was supposed to contain a liquid capable of giving life to hopeless invalids, was gazed at with deep emotion by many. I at once began the injections on a few selected cases at the Sanitarium, and watched the results with keenest interest. Koch had not at that time revealed the nature of his specific. Had I but known that the precious fluid was a glycerin extract of the tubercle bacillus I could have carried out my observations on a much larger scale, for in my little laboratory many flasks of liquid cultures of the tubercle bacillus were growing. . . . The bitter disappointment which followed within a few months the failure of Koch's treatment to bring about the miraculous cures which were expected from it, was shown very soon in a widespread and violent condemnation of the remedy, and for many years I had the utmost difficulty in obtaining the consent of patients and their physicians to the most cautious use of the injections. Nevertheless, so great was my faith in Koch, so convinced was I that whatever degree of immunization could be produced must be attainable by the poison of the germ or the germ itself, so impressed had I already become with what I had seen of the specific effect of tuberculin on animals and at the bedside, that I continued its study in the laboratory and its cautious use in patients who were willing to submit to the treatment. Thus it came about that through many long years during which the bitter prejudice against Koch's specific remedy continued unabated, tuberculin has been used continuously and cautiously at the Sanitarium ever since Dr. Osler's little vial of magic fluid reached Saranac Lake in 1890.[26]

Sherlock Holmes visits Robert Koch

The creator of Sherlock Holmes, the eminent writer A. Conan Doyle, was also a competent physician. He was sent to Berlin by the *Review of Reviews* to report on the Koch treatment, and in November 1890, he made the trip. A few passages arising from his visit have already been quoted elsewhere in this chapter, but additional quote gives insight into not only the tuberculin scene, but Koch's character at the time.

> . . . it is at present much easier to see the bacillus of Koch, than to catch even the most fleeting glimpse of its illustrious discoverer. His name is on every lip, his utterances are the constant subject of conversation, but, like the Veiled Prophet, he still remains unseen to any eyes save those of his own immediate coworkers and assistants. The stranger must content himself by looking up at the long grey walls of the Hygiene Museum in Kloster Strasse, and knowing that somewhere within them the great master mind is working, which is rapidly bringing under subjection those unruly tribes of deadly micro-organisms which are the last creatures in the organic world to submit to the sway of man.
>
> *The recluse of Kloster Strasse*
>
> The great bacteriologist is a man so devoted to his own particular line of work that all descriptions of him from other points of view must, in the main, be negative. Some five feet and a half in height, sturdily built, with brown hair fringing off to grey at the edges, he is a man whose appearance might be commonplace were it not for the vivacity of his expression and the quick decision of his manner. Of a thoroughly German type, with his earnest face, his high thoughtful forehead, and his slightly retroussé nose, he looks what he is, a student, a worker, and a philosopher. His eyes are small, grey, and searching, but so sorely tried by long years of microscopic work that they require the aid of the strongest glasses. A married man, and of a domestic turn of mind,[27] his life is spent either in the complete privacy of his family, or in the absorbing labour of his laboratory. He smokes little, drinks less, and leads so regular a life that he preserves his whole energy for the all-important mission to which he has devoted himself. One hobby he has, and only one . . . He is a keen

mountaineer, and never more happy than when, alpinstock in hand, he is breathing in the invigorating air of the higher Alps. Visitors at Pontresina last year may have observed there a quiet little sturdy gentleman, tweed-suited and be-spectacled, who vanished early from the hotel to reappear jaded and travel-stained in the evening; but few would have surmised that the energetic climber was none other than the renowned Professor of Berlin. . . .[28]

The invasion of Berlin by tubercular patients

As noted, over 1000 physicians flocked to Berlin to see the new treatment, accompanied by countless patients. Concern arose regarding the effect these patients would have on sanitation, especially since few of the patients could get into hospitals. The sanitation of the boarding houses and hotels where these patients were staying became an issue. The Police Department and the Board of Magistrates of the City of Berlin established emergency measures to control this problem. It was decided that regulations regarding disinfection would be applied not only to hospitals and clinics but to all places of a public character.

> The daily papers contain, as might be expected, reports of death among patients treated . . . by Koch's method. Such persons are not, unfortunately, endowed with immortality because they happen to have received an injection with Koch's liquid. So far no announcement has appeared . . . of any death in consequence of the use of Koch's liquid.[29]

Therapy or diagnosis?

Through late 1890 and early 1891 it gradually became clear that Koch's tuberculin was of dubious therapeutic value, but that its diagnostic value was inestimable. And it was, of course, as a diagnostic agent that tuberculin became the useful agent we know today. Dr. Koehler, an assistant of Bergmann at the Charité Hospital, had assumed responsibility for the treatment of surgical cases of tuberculosis. He summed up the views of most of the Berlin medical establishment at the end of 1890:

> 1. It is quite certain that Koch's liquid enables tubercular deposits to be discovered in the earliest stage, and in cases where their existence had entirely escaped notice. 2. It is a sure means of differential diagnosis in cases where other complicating diseases occur, which present the same or similar appearances to those found in tuberculosis. 3. Even though

the extreme benefits hoped for from Koch's method, especially by the general public, should not be derived, still its discovery must be regarded as one of the most important ever made in medicine.[30]

An extensive debate on Koch's remedy was also held during the winter of 1890–1891 at the Berlin Medical Society. (Koch did not attend.) This debate, reported extensively not only in the *Berliner Klinische Wochenschrift* but also in *The Lancet*,[31] involved all of the Berlin clinicians who were participating in tuberculin studies, as well as such scientific experts as Rudolf Virchow. The debate involved the detailed presentation of many cases, some of which had been putatively helped by tuberculin treatment. Finally, von Gossler ordered the heads of all Prussian university clinics to prepare a summary of all cases treated with Koch's remedy.[32] By this time, less than a year after Koch's initial announcement, 2172 persons had been treated with tuberculin, receiving a total of 17,500 injections. The forms of tuberculosis that were treated included tuberculosis of the lungs (phthisis), larynx, peritoneum, intestine, cerebral meninges, kidneys, bladder, lymph glands, and bones, as well as lupus, leprosy, pernicious anemia, pleurisy, and corneitis. Of the patients with incipient phthisis, 242 were treated, 9 were cured, 72 considerably improved, and 59 improved. Of those with moderately advanced phthisis, 444 were treated, 1 was cured, and 68 more or less improved. Of those with advanced phthisis (with cavities of the lungs), 230 were treated, 7 were considerably improved, 31 improved, and 30 died. Of lupus patients, 188 were treated, of whom 5 were cured, 78 considerably improved, and 84 less decidedly improved.[33]

Obviously, tuberculin was not a rousing success as a curative agent! Indeed, since there were no controls (the concept of the "double-blind" clinical trial was far in the future), it is difficult to know if *any* so-called "cure" was legitimate. However, Minister von Gossler was sufficiently convinced of the value of tuberculin that permission was granted for its use in prisons and in the Prussian army.[34]

How *was* tuberculin prepared?

Finally, in a paper published early in January 1891, Koch described the procedure by which tuberculin was prepared.[35] Why had he finally decided to publish? Was it because it was clear by now that tuberculin was *not* a "remedy"? We can glean a little of the background from Koch's words themselves:

As far as I can judge by the publications that have appeared ... my statements have, on the whole, found full confirmation. All agree that the remedy has a specific effect on tuberculous tissue and can therefore be used to search out hidden tuberculous processes that are difficult to diagnose. In regard to the therapeutic effect, most accounts also agree that, in spite of the relatively short duration of the treatment so far possible, many patients show improvement, varying only in degree. . . . I think the time has come for me to outline the steps by which I discovered the tuberculin phenomenon.

If a healthy guinea pig is inoculated with a pure culture of tubercle bacilli, the wound becomes rapidly healed during a few days. Only after 10 days to 2 weeks, a hard nodule forms, which soon opens, forming an ulcerating spot which persists until the death of the animal. On the other hand, if an animal that is already tuberculous is inoculated, the results are quite different. . . . In such cases, the small inoculation wound also becomes healed, but no nodule is formed. Instead, a peculiar change takes place at the point of inoculation. The spot becomes hard and dark-colored, and this region gradually spreads to a diameter of 0.5 to 1 centimeter. During the next few days it becomes more and more clear that the epidermis is necrotic. Finally, it is thrown off, and a flat ulcerated surface remains which generally heals quickly and completely. . . . Thus, the inoculated tubercle bacilli behave quite differently on the skin of a healthy guinea pig and one that is tuberculous. But this remarkable action does not occur only with living tubercle bacilli. Dead tubercle bacilli also bring about the reaction, whether killed by low temperatures of long duration, by boiling heat, or by certain chemicals.

It was then found that if pure cultures of tubercle bacilli are killed, ground up, and suspended in water, they can elicit the reaction. . . . These facts formed the foundation of my therapeutic method, because I could show that the curative effect on the tuberculous process could be obtained by a soluble substance which diffused into the fluids that surrounded the tubercle bacilli at the injection site. Thus, all that was left to do was to bring about the formation of these soluble substances *outside* the body—if possible, to extract and isolate the curative substance from the tubercle bacilli. This problem required much work and time before at last I succeeded, by the help of a 40 to 50 percent solution of glycerine. . . . *My new remedy against tuberculosis is therefore nothing more nor less than a glycerine extract of a pure culture of tubercle bacilli.*[36]

In actuality, Koch's discovery of this glycerine extract derived simply from the fact that glycerine was a component of the culture medium used to grow the tubercle bacilli. When the cell-free extract of the culture was concentrated by evaporation, the glycerine remained in the concentrate. In this paper, Koch gave very few details of how tuberculin was prepared, and these were finally given only in a subsequent paper.[37] [The details of tuberculin preparation need not concern us here.]

The origin of the name "tuberculin"

Initially, Koch's tuberculin had no real name. He first called it merely "remedy" (*Mittel* in German) but others began to call it names, some of them not too complimentary. During the latter part of 1890, the press often called it "Koch's lymph". Some physicians called it "Kochin", "Koch's fluid", or "bacillinum". Finally, Koch and Libbertz, in order to quell the actions of "amateur godfathers", designated the material "tuberculin". By February 1891, the labels of the bottles supplied read: "Tuberculin, Dr. Libbertz"[38] (Figure 18.2). Interestingly, a homeopathic remedy consisting of a dilution of tuberculous sputum had already been given the name "tuberculinum".[39]

Koch's new institute

We have already alluded to the new institute which von Gossler agreed to obtain for Koch, an institute which Koch asked for in order to pursue further his work on therapeutic measures for tuberculosis and other

Figure 18.2   *Tuberculin preparation made by culture on potato.*

diseases. Koch's desire to leave hygiene and work in therapeutics was motivated by his discovery of tuberculin, and it was the discovery of tuberculin that presented him with the *opportunity* of expanding his horizons. We have already seen how strongly he was supported by von Gossler. But the only insight we have into Koch's own mind at this time is a letter he wrote, near the end of 1890, to his former associate and favorite co-worker Georg Gaffky (who was at that time Professor of Hygiene at the University of Giessen). Koch says:

> I have often thought in recent months about how you and I had planned, a number of years ago, to look for a cure for tuberculosis. The first steps in this direction we took together. Then came a lot of interruptions, other duties, and now fate has taken us so far apart that I am unable to ask your help in the study of this new phenomenon we are working on.
>
> For me, the significance of the recent turn of events is that I am finally able to look forward to returning again to my first love, the study of infectious diseases. I have decided to leave the field of hygiene for good. The government has showed amazing willingness and speed in agreeing to establish for me an *Institut für Infektionskrankheiten*. Perhaps the whole thing will even be finished by April or May so that I can pull out for good from the Hygienic Institute on Klosterstrasse. But first I am going off on a vacation trip, something I need very much. I am going away with Dr. Schiess, who is here at the moment, on a month-long trip to Egypt. I hope nothing turns up to make me cancel this trip.[40]

And while Koch went off to Egypt, his son-in-law Dr. Pfuhl looked after the tuberculin studies and the preparations for the new institute. We discuss the new institute and Koch's further work on infectious diseases in the next chapter.

19

Consolidation and Transition

We are dealing here with a kind of experiment, an experiment which was performed on over 100,000 people, but which, in spite of its scale, conformed exactly with a laboratory experiment. In these two great populations, all the factors were the same except one, and that is the water supply. The population supplied with unfiltered water from the Elbe suffers seriously from cholera, whereas the population supplied with filtered water is virtually free of the disease. This difference is all the more important when it is realized that the water for Hamburg is taken from a place where the Elbe is relatively little contaminated, above the city, whereas Altona is forced to take water after it has received all the wastes of 800,000 people. Under these conditions, the scientist must conclude that the difference in incidence of cholera is due to differences in water supply, and that Altona was protected against cholera because it filtered its water.

—ROBERT KOCH[1]

In the winter of 1891, while the storm over tuberculin rose to a crescendo, Robert Koch left Berlin for an extended vacation in Egypt, leaving others to handle the problems which he had created. In the Berlin Medical Society, a three-month-long debate and discussion over the therapeutic value of tuberculin was in progress. The Prussian Landtag was arguing over the budget for Koch's new institute. Production of tuberculin was being pushed to a peak. And researches on tuberculosis and other diseases were being carried forward at the Hygiene Institute.

Meanwhile, Koch was sailing on the Nile together with friends he had made during his cholera days. His son-in-law, Eduard Pfuhl, held things together in Berlin and kept him informed about the progress on tuberculin and the new institute. There was deepening disappointment about tuberculin, and Koch answered bluntly:

> I was rather distressed to read your report on tuberculin. I'm not surprised that the demand for tuberculin has abated somewhat, considering

how badly it's being used in so many clinics. That all will change soon enough, I trust. . . . But what I am most concerned about now is the progress of the Institute for Infectious Diseases. You say in your letter that construction has gotten stalled because the Minister has not been able to get the Landtag to release the money. I plead with you to ask Herr Geheimrat Althoff about the status of the Institute, and if you get any real news, write me immediately, or perhaps even send me a telegram, so that I can change my travel plans if need be. I will not return to Berlin until this matter is cleared up. . . .

The trip has been fantastic, and I'm feeling rather well, even a little tanned from the Egyptian sun. My best wishes to all, and especially to Libbertz.[2]

The problem about the institute was apparently not the money for construction, since the physical work was already underway,[3] but the operating budget.

One can look at Koch's request for a new institute in one of two ways: 1) He genuinely believed that the time was ripe for a real research program on the therapy of infectious diseases and concluded that the only way to get it was to have a research institute allied with a hospital, so that clinical and laboratory research could be integrated; or 2) He was bored with his professorial tasks, found his Hygiene Institute, situated in an old building, unsatisfactory, and his duties onerous, and used the tuberculin excitement as an excuse to get a new institute and a new building. The Pasteur Institute had just been completed in Paris, and the competitive spirit he felt toward Pasteur may have driven him toward a desire for a similar institute.

The plans for the Institute of Infectious Diseases

As discussed in Chapter 18, Koch and the Minister of Education had agreed to the establishment of a new institute already in October of 1890, and the plans were completed on 6 November 1890. The plan was that the institute should be in two parts, a hospital for clinical testing, and a laboratory for scientific research. The Pasteur Institute in Paris served in part as a model. For financial reasons, the hospital was to be built on land near the Charité Hospital, with a total of 128 beds in four separate wings. The scientific department was to be nearby. The institute was organized in such a way that the overall direction would be Koch's, but under him there would be two directors, one for the clinical, the other for the scientific, departments. The first clinical director was Ludwig Brieger (1849–1919)[4] and the first scientific director was Richard Pfeiffer. The scientific department was to have space for 20 independent research workers. Koch himself would be relieved of all administrative duties, thus being free to carry on his own independent research.

The Charité Hospital

The first Charité Hospital was developed out of a building constructed outside the walls of Berlin in 1710 for the treatment of smallpox patients. In 1785–1797 the side wings were added and in 1800 the middle part was reconstructed (see Figure 3.1). In 1831–1834 the so-called "New Charité" was built, originally for smallpox patients, and additions were subsequently made in 1851, 1856, and 1866–67. The Pathological Institute associated with the Charité complex was erected for Rudolf Virchow (as one of his conditions for returning to Berlin from Würzburg[5]) in 1856 and significantly enlarged in 1873. With the rapid developments in medicine after the establishment of the German Empire, in the 1870s many additional buildings were added, so that by the time planning for Koch's institute began, there was a veritable warren of buildings—some old, some new, some built especially for medicine, others rebuilt from apartment buildings.[6]

In the fall of 1890, construction for the Institute of Infectious Diseases began. In spite of great difficulties as a result of the cold winter, the building was completed and was being occupied by July 1891.[7]

The buildings of Koch's new Institute

Koch's new clinical facilities occupied the land between the Old Charité and the Stadtbahn (elevated railroad) (Figure 19.1; see also map, Figure 9.1) and consisted of a series of small wooden buildings (called by the Germans "barracks"). A single large clinic could not be built because of the shape of the available land and the unfavorable subsurface terrain. By constructing a series of small buildings, individual groups of patients could be kept totally separate. The whole arrangement was designed according to the new "reformist" movement for hospital care, with easy access to light and air. Special care was taken in the prevention of hospital infections (*Hausinfektionen* in German), showing an early concern for nosocomial infections.[8]

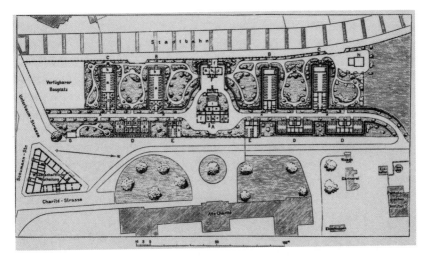

Figure 19.1   *Plan of the Koch Institute of Infectious Diseases in Berlin (see Figure 9.1 for location). The Triangle building is on the left, the "barracks" in the upper center (labeled B, C, and D). The autopsy and disinfection building is labeled F. Quarters for attendants are labeled E. The original Charité Hospital is in the bottom center and Virchow's Pathological Institute is in the lower right corner. The S Bahn (see Figure 9.1) runs across the top of the figure.*

The research laboratory was established in an old three-story apartment house which the Charité adminstration had purchased a few years before. This building, the so-called "Triangle", was situated where three roads came together (Figure 19.2). The large number of windows in this building appealed to Koch, since it made the building very suitable for microscopy. [Although electricity was beginning to be used in 1891, it would be a few years yet before artificial lighting would be used for microscopy.] The Triangle was much larger than the Hygiene Institute on Klosterstrasse and had extensive office and laboratory facilities. Koch had his own private suite of offices and laboratory on the second floor, separate from those of the other workers (Figure 19.3). An examination of the plans of this building[9] shows that the rooms were rather small (Figure 19.4) and the communication between rooms was difficult. Koch's office was 4.6 meters square and his laboratory 3.36 meters square.[10]

The operating budget for the Institute was debated strenuously by the Prussian Parliament on 9 May 1891.[11] During the debate, one of the members of parliament stated that although: "the therapeutic value of tuberculin was exceedingly doubtful, its scientific importance was of the very highest order. . . . It is the duty of Germany to accede to the

(a)

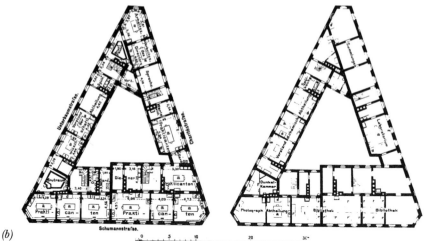

(b)

Figure 19.2 *The Triangle laboratory. (a) Photograph of the building on Schumannstrasse. The building was destroyed during World War II and the street on which it was has been altered. (b) Floor plan of the Triangle laboratory. The second floor is diagrammed on the left, the third floor on the right. Schumannstrasse is at the bottom of the figure, Charitéstrasse to the right. Robert Koch's laboratory and office were in the rooms in the upper part of the second floor triangle. Most of the rest of the rooms on this floor housed assistants (Prakitcanten). The third floor housed the library (Bibliothek), photographic department and darkrooms, chemical laboratory, and culture collections (Sammlungen).*

Figure 19.3   *Robert Koch's office in the Triangle laboratory.*

demands of the Budget Commission." An opposition member raised again the strong objections about research with secret remedies: "I quite agree that the State should do all in its power to secure the continuation of these investigations. But there must be no more secrecy; there must be no more experimenting on human bodies with a secret remedy." Herr Geheimrat Althoff, speaking for the Government, responded: "The institution will not be used for experiments with a secret remedy . . . all the scientific work and the discoveries that might be made within its walls would be freely published. . . . The whole medical world is unanimous that we are on the threshold of a new therapeutic era, and the fight against infectious diseases must be taken up with renewed vigor." Rudolf Virchow, who was a member of the Parliament, supported the institute, although emphasizing that tuberculin was only one of many things that would be studied there. However, he rather pointedly noted that Koch's suggested salary was enormous, and even the salaries of Koch's assistants were to be higher than those of many German professors. Virchow concluded: "I may remark that the annual expenses represent a sum which is about equal to that received by all the scientific departments of the University together for purposes of research." After a few more remarks, the debate was closed and the requested sum approved.

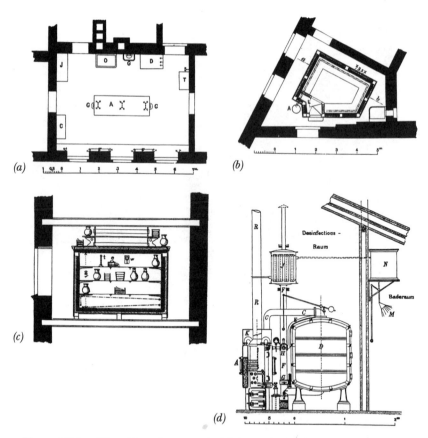

Figure 19.4   *Floor plan of some typical laboratories in the Triangle building. (a) Laboratory. Key: A, work table; C, chemical cabinet; D, digestion apparatus; E, wall cabinet; G, sink; J, instrument cabinet; O, oven; T, table. Gas outlets are indicated by symbols. (b) Plan of an incubator room. The temperature of the room was controlled by hot water (route of pipes shown in dotted lines around the edge of the room). The walls were heavily insulated (x, wood panelling; y, stone; z, gypsum panelling). (c) Plan of an incubator. The interior height is 1.80 meters. (d) Plan of a disinfection apparatus. The large autoclave is shown in the center (D). Outside the disinfection room is a wash room with shower.*

Later, it was reported that Koch's salary was to be 20,000 Marks, which was more than 10 percent of the whole budget of the institute.[12]

Adminstrative changes

When Koch asked for the establishment of the new institute, he also resigned as director of the Hygiene Institute on Klosterstrasse, as well as resigning his professorship at the University. As noted in the last chapter, he stated that his primary motivation for resigning his professorship was so he would have time to pursue his tuberculin research. He was appointed an honorary professor, which meant that he was free to do research as he pleased, but was no longer required to give his regular course of lectures. Koch was naturally interested in having one of his own people replace him as Professor of Hygiene and Director of the Hygiene Institute. He earnestly requested that Carl Flügge be appointed to replace him. Flügge, who was one of Koch's most intimate colleagues (Flügge and Koch had been the co-founders of the *Zeitschrift für Hygiene*), was at that time Professor of Hygiene at the University of Breslau. Unfortunately, the Minister of Education did not accede to Koch's request, appointing instead Max Rubner (1854–1932), a follower of the Pettenkofer school of hygiene. Koch himself announced the news of Rubner's appointment to Flügge, and we can appreciate the university politics that must have been involved in the decision:

> All our hopes are shattered. Geheimrat Althoff has just told me that the decision on my successor has been made and that it will be Rubner. He will come in the next few days to talk with me about taking over the Institute. I am absolutely destroyed by the news, since I had assumed all along that the position would go to you. . . . If I had more influence with the Faculty, it might have gone differently, but the loud voices in the Faculty now are against me, as you might expect. In the last weeks I had some indications of what was going to happen but I was so stupid that I calmly let the whole thing get out of hand, thinking that I had justice on my side. Perhaps it's a tiny consolation for you to know that you weren't the only one affected by this folly.[13]

The Medical Faculty, who had the final say in Koch's replacement, had fallen under the influence of Virchow and another professor, Oskar Liebrich, who had claimed to have found his own cure for tuberculosis. They opted for a legitimate hygienist of the Pettenkofer school rather than a bacteriologist.[14]

The work of the Institute

The "Triangle" thus became the research headquarters of Robert Koch during the years 1891–1904. Into this building he moved with a host of eager assistants and co-workers, bringing with them the research programs that had been underway on Klosterstrasse. As noted, there was room for 20 assistants, and with the handsome operating budget, the place was soon humming with activity. Among those who worked in the Triangle, the most noteworthy were Emil Behring and Shibasaburo Kitasato (see below), Paul Ehrlich, Richard Pfeiffer, Bernhard Proskauer (1851–1915), and August von Wassermann (1866–1925), all names well known in the history of bacteriology. (Wassermann, the discoverer of the famous Wassermann test for syphilis, had a checkered career in Koch's laboratory. He came to Koch's institute in 1891 as a 25-year old, but Koch refused to give him a paid postion. Wassermann worked for *10 years* as an unpaid assistant before he finally was appointed as a section leader! He remained in the institute for 22 years before being appointed in 1913 as the leader of a new Kaiser-Wilhelm-Institut für experimentelle Therapie.)

Pfeiffer (Figure 19.5), the scientific director of the Institute, became

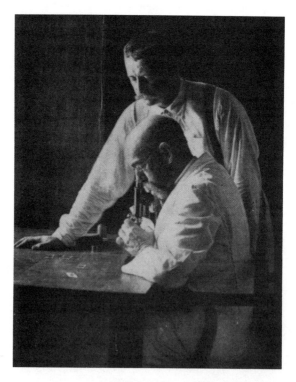

Figure 19.5   *Robert Koch and Richard Pfeiffer in the Institute for Infectious Diseases, 1892.*

well known for his discovery of immune lysis (the "Pfeiffer phenomenon"), a concept which became one of the major entrées into the emerging field of immunology (and which incidentally helped to unravel the cholera epidemic of Hamburg a year later). Behring and Kitasato discovered antitoxin (see below), and Ehrlich became a leader in both immunology and chemotherapy.[15] Proskauer, the head of the chemical department of the institute, is best known for his participation in the development of the Voges-Proskauer test, one of the main diagnostic tests for enteric bacteria.

Koch himself continued work on a variety of problems, including his work on tuberculin. But an outbreak of cholera led the institute back into further research on this dread disease. However, before discussing some of these research programs that Koch himself was involved in, we must digress a little to discuss Behring's famous work.

Behring's work on diphtheria antitoxin[16]

Emil von Behring (Figure 19.6), one of Koch's most important collaborators, and, later, competitors, was born in 1854, eleven years after Koch. He was from a poor family with a large number of children and he was able to go through medical school only because he joined the army. He completed his medical training at the Charité in 1878, the year that Robert Koch was publishing his noteworthy book on wound infections. After medical school, Behring became an army doctor in the district of Posen, where he worked from 1880–1883. He had a strong interest in research and while in Posen studied the use of iodoform as an antibacterial agent. He also studied the antibacterial effect of serum on *Bacillus anthracis*, making the interesting discovery that guinea pigs were highly sensitive to this organism and that rats were resistant. He was able to correlate the resistance in rats with an inability of the organism to grow in rat serum. Metchnikoff had recently published his first work on phagocytosis but Behring could easily show that the inability of *B. anthracis* cells to grow in rat serum was not due to the presence of phagocytes, but to the effect of a soluble factor.

Behring transferred back to Berlin in 1888 as a lecturer in the Army Medical College and in 1889 was assigned to work in Koch's institute on Klosterstrasse. Sibasaburo Kitasato (1852–1931) was also at Koch's institute at this time, working on tetanus (Figure 19.6*b*). When Behring first went to Koch's institute, he worked with F. Nissen on antibacterial

Figure 19.6   *Behring and Kitasato. (a) Emil von Behring (right) and an assistant. (b) Shibasaburo Kitasato with his apparatus for growing bacteria under anaerobic conditions.*

substances in serum,[17] and subsequently began to work with Kitasato on toxins. Kitasato had isolated the causal agent of tetanus, now known as *Clostridium tetani*, in 1889, and Kitasato and Weyl, following Brieger's work, showed that the organism produced in culture a powerful toxin.

Work was also initiated on diphtheria. Some years before, Loeffler had isolated the causal agent of diphtheria, now known as *Corynebacterium diphtheriae*, and in 1889 Émile Roux and Alexandre Yersin had shown that its pathogenicity was due to a toxin.

Then came the key demonstration that animals which had become immune to diphtheria and tetanus produced a substance which neutralized the respective toxins.

Pfeiffer had just demonstrated a factor in serum that was able to bring about the lysis of *Vibrio* cells and at this time Pfeiffer's bacteriolysis and Metchnikoff's phagocytosis were considered to be the only possibilities for immunity. The role of toxins in pathogenesis was a new concept, and the discovery that a soluble factor in immune serum *neutralized* a toxin was a major finding. Because of the powerful nature of the tetanus toxin, the experiments were most clearly done with this system, and it is probably for this reason that the first paper by Behring and Kitasato deals with tetanus.[18]

> 1. The blood of rabbits immune to tetanus has the ability to neutralize or destroy the tetanus toxin. 2. This property exists also in extravascular blood and in cell-free serum. 3. This property is so stable that it remains effective even in the body of other animals, so that it is possible, through blood or serum transfusions, to achieve an outstanding therapeutic effect. 4. The property which destroys tetanus toxin does not exist in the blood of animals which are not immune to tetanus, and when one incorporates tetanus toxin into nonimmune animals, the toxin can be still demonstrated in the blood and other body fluids of the animal, even after its death.[19]

The science of serology can be said to have begun with this paper. Of even greater importance, the discovery opened up the possibility of *therapy*, especially for diphtheria, which at that time was the major infectious disease of children. By injecting serum containing the neutralizing factor prepared in another animal, it seemed possible that the effects of the toxin might be alleviated (passive immunity), thus giving time for the host's own defenses to take over. In addition, if the antitoxic agent could be induced in a host before infection occurred, then an active immunization procedure was possible. The first use of the term *antitoxin* occurred in a footnote to the Behring/Kitasato paper, and the paper concluded with the challenging phrase from Goethe's Faust:

> Blut is ein ganz besonderer Saft. (Blood is a very unusual fluid.)[20]

As we discussed in Chapter 18, Koch's work on tuberculin and the work of Behring and Kitasato on antitoxin were published in the same year. Koch's discovery provided the basis for the field of cellular immunology and Behring and Kitasato's for that of humoral immunology. It is amazing to consider that both these discoveries were made in nearby laboratories in the same building on Klosterstrasse. Although Koch's discovery caused the most immediate excitement, it failed to live up to its promise. But it quickly became clear that the work of Behring and Kitasato was of immense practical significance. Lister alluded to this work in the report on his visit to Koch's laboratory in the fall of 1890 (at a time when the work had not yet been published, see Chapter 18).

One week after the Behring and Kitasato paper, which dealt primarily with tetanus but alluded to diphtheria, Behring alone published his paper on diphtheria.[21]

Kitasato went back to Japan in 1892, where he founded a bacteriological institute (now the Kitasato Institute in Tokyo). Behring moved with Koch to the new Institute for Infectious Diseases at the Charité and continued to work on diphtheria antitoxin. In 1892, Behring and Wernicke published a paper describing the use of diphtheria antitoxin in the *treatment* of experimental diphtheria in animals,[22] work done at the Triangle laboratory. In the same year, Behring wrote several lengthy papers on serum therapy that were privately published. Although this work was all done in Koch's institute, much of it was privately financed by Behring himself, including the purchase of the animals needed for the *in vivo* experiments. Behring credited Koch for advice and support ("During the whole period of the experiments, Geheimrat Koch has given us his advice and counsel") but when a large amount of money was needed to commercialize the discovery, Behring turned to industry. Koch and Althoff, the head of the Department of Education at that time, suggested that Behring work with the dye company *Meister Lucius and Brüning* in Hoechst, a suburb of Frankfurt. (This company became the forerunner of the great Farbwerke Hoechst pharmaceutical firm.) Behring entered into a contract with the dye company for the production on an industrial scale of diphtheria antiserum, using dairy cattle. The first clinical trials were carried out in Munich, Leipzig, and Berlin at the end of 1892. Despite a number of problems, especially with low potency of the preparations and lack of adequate standardization, the first trials showed promise of success.

At the same time, Émile Roux in France was carrying out his own studies with Behring's antitoxin, using preparations that had been made in horses. The horse turned out to be a better animal for producing high titer antitoxin than the cow, and Roux had great success in clinical trials. In the Hôpital des Enfants Malades in Paris, the death rate due to diphtheria dropped from 51.7 percent to 24.5 percent after the introduction of Roux's antitoxin preparation. Roux's success led to another Franco-Prussian altercation, with the French press claiming that Roux was the discoverer of antitoxin therapy. Roux himself vigorously denied this claim, and Roux and Behring subsequently shared a prize of the Académie des Sciences of France, Behring as the discoverer of the new remedy and Roux as the first to introduce it into France. Subsequently, Behring alone received a prize from the President of the French Republic for his work [and the first Nobel Prize in Medicine, in 1901].

Paul Ehrlich

All sources seem to agree that if it had not been for Paul Ehrlich, Behring's honors would have come to naught. For it was Ehrlich who first developed suitable quantitative methods for diphtheria antitoxin, thus permitting analysis of diphtheria antitoxin production and efficacy.

We have met Paul Ehrlich before in this book, since it was Ehrlich who perfected Koch's method for staining tubercle bacilli (see Chapter 14). Ehrlich, born in Silesia of a prosperous Jewish family, studied medicine and pathology at the University of Breslau under Carl Weigert and Julius Cohnheim. It was in Breslau that Ehrlich first met Robert Koch, during Koch's fabulous Wollstein days:

> One day a physician from the town of Wollstein in Silesia came to Breslau to demonstrate to the University professors there his investigations on the anthrax bacillus. The guest was shown through the different laboratories and was also taken to the bench where the student Ehrlich was working. Watching him, the guest was told:
> "That is 'little Ehrlich'. He is very good at staining, but he will *never* pass his examinations."
> This visiting physician from Wollstein was none other than Robert Koch, whose name was soon to become famous throughout the world. In later years he was very often in contact with this same "little Ehrlich" and worked successfully in close collaboration with him.[23]

Ehrlich completed his thesis in 1878 and went on to Berlin, where he became a physician and later Senior House Physician at the Charité. It was while he was working at the Charité that Ehrlich attended Koch's famous lecture at the Physiological Society on the tubercle bacillus, and overnight improved on Koch's staining method (see Chapter 14).

A year or so later, Ehrlich himself showed symptoms of pulmonary tuberculosis. Unhappy with his position at the Charité because it was not offering him sufficient time for research, and with his tuberculosis getting worse, he gave up his position in Berlin and went to Egypt, hoping that the warm dry air would bring about a cure. He stayed in Egypt for two years, returning in 1889 strong and healthy. Now, using his modest personal wealth, he set up his own little private laboratory in Berlin and carried out simple researches at his own expense, primarily on staining methods. When Koch moved into his new Institute of Infectious Diseases, he offered Ehrlich a position, which Ehrlich happily accepted. Behring was now in the middle of his important studies on diphtheria antitoxin and Ehrlich began to work with him. Soon, a close friendship developed between these two workers.

As noted, Behring was attempting to produce diphtheria antitoxin on a large scale for clinical trials. However, he was having much difficulty obtaining potent preparations and the research was threatened with complete failure. Among his other research, Ehrlich had done extensive work on antisera prepared against the plant toxins *ricin* and *abrin*, and had developed quantitative procedures for assaying the potencies of these antitoxins. At this point, Ehrlich began to give Behring the benefit of his quantitative immunological experience. Using experimental animals, he worked out a precise method of measuring the potency of antitoxin in units defined in relation to a fixed standard, probably the first biological standard ever developed. [The standards which Ehrlich set up are still the basis for measuring diphtheria antitoxin as used throughout the world.] Using his standard and a quantitative assay method, Ehrlich produced a potent diphtheria antitoxin in animals by means of repeated injections.

Now that high-potency antitoxin was available, its use as a remedy in the treatment of diphtheria could begin in earnest. On 14 October 1893, Ehrlich and Behring signed a written agreement regarding the commercialization of diphtheria antitoxin. However, they had a falling out soon afterward and Behring developed the diphtheria antitoxin independently. Ehrlich continued to work at Koch's institute until 1896, when he was made Director of the Serum Institute at Steglitz (a suburb of Berlin). In this position, he was responsible for the state approval of all sera, and Behring's antitoxin had to pass through his control. Behring received much acclaim for his discovery of antitoxin, and was the recipient of the first Nobel Prize in Medicine (in 1901). He continued work on immunization against infectious disease and eventually developed a procedure for active immunization against diphtheria using toxoid.[24] Ehrlich himself went on to make major contributions to immunology and chemotherapy and was awarded the Nobel Prize in 1908. He remained bitter about how he had been treated by Behring, however, and in 1899 wrote a friend:

> I always get wild whenever I think of that dark period and the way in which B. tried to hide our scientific partnership. But the revenge has come. He can see how far he has got without me since our separation. Everything is blocked now: his work on plague, cholera, glanders, streptococcal infections. He makes no progress with diphtheria . . . And all this with more than sufficient means in hand, and a swarm of collaborators–Ruppel, Lingelsheim, Knorr, Japanese workers, and assistants of the big factory.
>
> Of course, you can imagine how filled with rage he is. He wanted to be the "All-Highest" who would dictate his laws to the entire world and who, in addition, could earn the most money. He wanted to be a Superman; but—thank God—he did not have the necessary super-brain. . . . Away with the mammonisation of science![25]

The new work on cholera

While Behring and Ehrlich were scrapping over diphtheria antitoxin in Robert Koch's institute, Koch himself was not idle. In August 1892 there was suddenly a new outbreak of cholera in Hamburg, and Koch was pulled away from his studies on tuberculin in order to pursue it. We can imagine that Koch might have welcomed a respite from tuberculin, now that its therapeutic value was increasingly in doubt, but the Hamburg epidemic was very serious. In several weeks, over 17,000 cholera cases were reported in the Hamburg area, with over 8,000 deaths. The Senate of Hamburg called for assistance, and Koch threw all the resources of his laboratory into the fray. Georg Gaffky was brought up from Giessen to study the outbreak on the spot. This was the first substantial cholera outbreak in Germany since Koch's discovery of the causal agent, and the knowledge gained about the bacterium and the disease in the ten years since Koch's discovery could be used to develop more effective public health measures.

At the time of this outbreak, the Pettenkofer school of hygiene was still insisting that cholera arose primarily because of atmospheric or terrestrial influences. The role of water, if any, was considered to be only secondary, perhaps modifying the susceptibility of the host. Koch in particular was strongly attacked for his insistence on cholera as due to an infectious agent, and water as the primary mode of transmission.

The Hamburg epidemic permitted an unequivocal proof that cholera was waterborne. Even more, it permitted Koch to draw a direct relationship between the *purity* of the water and the incidence of cholera. In this very influential work, which had far-reaching importance for the public health of an increasingly urbanized population, Koch showed that *water filtration* was the key to an effective attack on the cholera problem. Koch's work, published in a paper in his own journal, is a model of clarity and logic. We see the master, in this paper, back in true form.[26]

How did Koch deduce the importance of water filtration? Two cities, Hamburg and its neighbor Altona, obtained water from the Elbe River, but Hamburg obtained its water unfiltered from *above* the city, whereas Altona obtained its water from *below* the city of Hamburg. Because Altona's water was derived from a much poorer source than Hamburg's (the water had flowed through the city of Hamburg, picking up sewage on the way), Altona filtered its water supply, using slow sand filters, whereas Hamburg used its (presumably better) source water unfiltered. But while Hamburg was visited heavily with cholera, Altona was nearly free of the disease. Koch's paper lays out the situation:

> Most surprising was the situation at the border between Hamburg and Altona. On each side of the border, the conditions are identical: soil, construction of the buildings, sewerage, population. In short, all of the environmental factors are identical, yet cholera is found right up to the border of Hamburg, but is not found at all in Altona. In one group of houses, in an area called Hamburger Platz, the incidence of cholera marks out the boundary of the border between the two cities better than any map could have done. Cholera traces not only the political boundary between the two cities, but even the boundary of the water supply, because here there is a group of houses which belongs to Hamburg yet obtains its water from Altona. These houses remain completely free of cholera, whereas all around in the rest of Hamburg there are numerous cases of the disease. We are dealing here with a kind of experiment, an experiment which was performed on over 100,000 people, but which, in spite of its scale, conformed exactly with a laboratory experiment. In these two great populations, all the factors are the same except one, and that is the water supply. The population supplied with unfiltered water from the Elbe suffers seriously from cholera, whereas the population supplied with filtered water is virtually free of the disease. This difference is all the more im-

portant when it is realized that the water for Hamburg is taken from a place where the Elbe is relatively little contaminated, above the city, whereas Altona is forced to take water after it has received all the wastes of 800,000 people. Under these conditions, the scientist must conclude that the difference in incidence of cholera is due to differences in water supply, and that Altona is protected against cholera because it filtered its water. . . .

For a bacteriologist, nothing is easier than to explain why cholera is restricted to Hamburg. The cholera bacteria are brought into the Hamburg water either from the Hamburg sewers, or from the dejecta of persons living on the boats anchored near where the water is taken. . . . Altona takes its water from a source which is much worse than Hamburg's, but careful filtration renders it completely, or nearly completely, free of cholera bacteria. . . .

[And then a dig at Pettenkofer]:

Why one would wish to attribute the Hamburg-Altona cholera to cosmic, telluric, or purely meteorological factors is a puzzle to me. The sky, sun, wind, rains, etc. are absolutely equally distributed on the two sides of the city boundary.[27]

Koch then proceeded to provide solid bacteriological evidence for the efficacy of sand filtration. Bacterial counts were made of water before and after filtration, which showed that filtration was highly effective in removing bacteria. Although it was later found that filtration alone was not sufficient to provide a consistently safe drinking water (chlorination was introduced about 1910), the demonstration by bacteriological means that filtration was important provided the *methodology* which was needed by engineers in order to perfect proper water supply systems. Koch's bacteriology made it possible to place sanitary engineering on a firm footing.

This work was so clear and convincing that it became the basis for regulations prepared by a government commission in which Koch participated.[28] These regulations required the bacteriological examination of all public water supplies in Germany which used surface water, and prescribed exactly how the filters should be evaluated. These regulations became the first standard methods for the examination of water, and served as the forerunner for all later standard methods, including those of the English-speaking world.

It is of interest that Koch had wanted to leave work on hygiene to return to infectious disease, but he was barely established in his new institute before he was involved in one of the most important hygiene issues of the day. Circumstances, of course, had pulled him into the cholera study, but he may have lost some of his "enthusiasm" for tuberculosis and tuberculin by this time.

Another important issue that arose out of the cholera work in Hamburg was the separation of cholera-like bacteria from the true cholera vibrio. Vibrios are, of course, widespread in waters, and nonpathogenic vibrios are readily isolated from lakes and streams. The separation of these nonpathogens from the cholera vibrio was carried out by immunological means, the first such use in bacteriology. As noted earlier, Koch's co-worker Richard Pfeiffer had discovered the phenomenon of immune lysis, observing the rapid lysis of vibrios by specific antisera. Antisera were prepared against a variety of vibrios and the antiserum specificity provided a means for distinguishing pathogenic from nonpathogenic types. This work, one of the first solid pieces of bacteriological research from Koch's new institute, set the spirit and tone of the young institute.

Changes in Koch's personal life

Robert Koch had not been on good terms with his wife for a number of years (Figure 19.7a). About the time that the tuberculin crisis broke, Koch became acquainted with a young art student, Hedwig Freiberg (Figure 19.7b). Hedwig, a pretty girl of 17, awoke in Koch a passion that had been long dormant. By this time, Koch's daughter Trudy had

(a)

(b)

Figure 19.7 (a) Emmy Koch in 1891. (b) Hedwig Freiburg Koch. The inscription over the picture can be translated: "As I was when I first met Robert Koch, 1889." Both photographs courtesy of the Robert Koch Institute, West Berlin.

married and left home, and Koch had become increasingly morose over how the tuberculin discovery had affected his scientific life. Although he did not marry Hedwig until three years after they met, Koch was already intimate with her when he went off to Egypt in the winter of 1891. In a letter which was subsequently published in facsimile form,[29] Koch poured his heart out to this nubile youngster:

> My dearest Hedchen!
> I have had an absolutely wonderful time in Luxor. What a time, with the blue heavens laughing over the green fields of the upper Nile . . . I've been climbing up to places that only eagles reach, and I have gone far out in the desert where only the bedouins travel. And over all are ruins of old temples, ancient paintings with inscriptions, tombs with relics, which my friend Kartulis and I are hunting all over this old land of the Pharaohs. I could have stayed for months in Luxor, but I was pulled like a magnet north to Cairo, where I hoped to find a sweet letter from you. But nothing! Instead of that I found a letter from Berlin with a less pleasant message. You know that you are my heart and soul, and now I wait heartbroken for news of my Hedchen. You must know that my discovery [of tuberculin] has brought out the vultures. Virchow, of course, who works on all the remedies, but in addition, Prof. Liebreich claims to have found a remedy which is better than mine. Above all I believe that my work will be successful, but it's going to take a bit of time to prove it. At the moment successes with tuberculin are few and far between, and not very much is being sold. But for me, the most important thing is getting the money to build my institute. Even this is now in question. Before April 1 the Landtag will decide if they are going to appropriate the money, and until that happens I can't come back to Berlin. I must stay away because it would be embarassing if such a delicate question was under discussion in my presence, but I fear that the Minister is dragging his feet. So at the moment everything is in a bad way, but don't lose your hope. I'm going to hold fast to my conditions, come what may. But in the meantime, please write! Tell me what you are thinking and if you are thinking badly of me. I'm going to stay in Cairo until the middle of the month, and then on to Alexandria, where I will await the decision of the Landtag. As soon as that is settled, I will hurry back to Berlin. Dearest Hedchen, if you love me, then I can put up with anything, even failure. Don't leave me now, your love is my comfort and the beacon that guides my path.
> Heartfelt greetings and kisses come to you from your own, your ever faithful
>
> Robert[30]

Finally, in 1893, Emmy Koch went back to Clausthal, leaving Robert to his paramour. Koch had purchased his childhood house (see Figure 2.2c) in 1891 and now, as part of the divorce settlement, he gave it to

Emmy Koch. In this house, Emmy lived out her life, welcoming her daughter, her son-in-law, and her three grandchildren. The Pfuhls were now resident in Strassburg, where Eduard Pfuhl had been transferred in the winter of 1892/93.

Robert Koch and Hedwig Freiberg were married in Berlin on 13 September 1893. Robert was almost 50 years old, Hedwig 20. Hedwig became Koch's constant companion, especially on his increasingly frequent and strenuous trips abroad (see Chapter 20). After Koch died in 1910, Hedwig Koch traveled extensively in the Far East, especially to study the religions of Japan and China. She died in Germany 16 June 1945.[31]

Further work on tuberculin

In the decade of the 1890s, at his new Institute, Koch's other major research had to deal with the preparation and testing of tuberculin. Although most people were convinced by now that tuberculin was not of therapeutic value, Koch felt obliged to persist in this work. However, as time went on, the main thrust of the work was toward an understanding of the immunology of tuberculosis, with the goal of vaccine development. Behring, who had by this time left Koch's institute for a position in Halle (and later in Marburg), began himself to work on tuberculosis, basing his research program on the concept of the toxin-antitoxin interactions that he had developed for diphtheria. And gradually, Behring's work developed into a controversy with Robert Koch.

Koch himself was eager to improve tuberculin. Much work was directed at the proper procedures for growing the bacteria and extracting the active agent. By now, the aim was not to *cure* the disease but to *immunize* against it. It was clear that the tuberculin reaction that Koch had first described was an immune reaction, albeit complex and ill-defined. Behring had developed an active immunization procedure for diphtheria, and it seemed reasonable that this should also be possible for tuberculosis. All one needed to do was find the right material. By 1897 Koch was writing: "Immunity against some infection diseases can indeed be easily achieved, but it is not necessarily simple and might require the joint participation of two or more different components."[32] In the case of tetanus and diphtheria, immunity could be achieved by the simple approach of neutralizing the toxin. In the case of cholera and typhoid, Pfeiffer had shown that immunity was directed against the cells rather than against a toxin. Koch felt that the ideal type of immunization procedure would be one which was not only against the toxin, but against the organism as well.

Unfortunately, immunity to tuberculosis is not a simple problem. Koch had clearly shown, even in the first days of his tuberculin work, that tuberculin itself had no effect against the bacterium, but only against diseased tissue. Indeed, in miliary tuberculosis (and in experimental infection in the guinea pig, which seemed to mimic miliary tuberculosis in humans), the bacteria were very invasive and immunity developed slowly or not at all. This would suggest that artificial immunization against tuberculosis would be difficult to achieve and that more than one material might be necessary.

Following this logic, Koch came to the decision that more than one type of agent would have to be developed. In addition to his original tuberculin, he tried to achieve immunity by injecting preparations of killed tubercle bacilli, following this with increasing doses of living cultures. In this approach, he was, of course, following the road that Pasteur had first taken in his work on rabies. This work was done by Koch's assistants Fred Neufeld and August von Wassermann in goats, donkeys, and cattle, testing for the success of the immunization by subsequently injecting virulent tubercle bacilli.

However, experiments using dead bacteria as the immunizing agent were a failure. Numerous other experiments followed, all directed at an attempt to achieve immunity by means of tubercle bacilli or their products. Koch finally concluded that tuberculosis was not a disease in which immunity could be achieved by use of the bacteria themselves. He carefully contrasted tuberculosis with the other bacterial diseases where such immunity could be achieved: typhoid fever, cholera, and plague.

Meanwhile, Behring was not idle. Having moved to Marburg in 1895, Behring was fresh from his triumph with diphtheria and ready to move on to other things. With his student Paul Römer, Behring had begun extensive immunization experiments using living tubercle bacilli. The details of this work are of little current scientific interest, but resulted in strained relationships between Behring and Koch for the rest of Koch's life. Unable to develop an effective serum therapy for tuberculosis, Behring had turned to vaccination studies on cattle, developing a succession of vaccines that were of dubious value. Koch, the discoverer of the tubercle bacillus and tuberculin, considered Behring an interloper in the tuberculosis field, and strongly resented Behring's activities.

The conflict between Behring and Koch came to a head when Behring and the Farbwerke Hoechst chemical company obtained patents for two different extracts of tubercle bacilli, despite vigorous opposition from Koch. However, Behring's preparations proved not to be useful.

The other side of the tuberculin story, the use of the agent as a diagnostic tool, has had immense practical significance. However, the initial use of tuberculin for diagnosis was by veterinarians in dairy cows. But even tuberculin diagnosis in dairy cattle became embroiled in a rather extended controversy between Koch and others about the identity of the cattle and human strains of tubercle bacilli. We discuss this controversy in Chapter 21.[33]

Other activities

Koch's work in the cholera emergency in Hamburg brought him increasing public responsibility. In addition to serving on a wide variety of government commissions regarding public health matters, he arranged for the establishment at his Institute of immunization stations for the Pasteur treatment of rabies, and for standardization of diphtheria antitoxin (this latter was taken over by Paul Ehrlich when he became head of the State Serum Institute). Communicable disease laws promulgated by the German government had Koch's imprimatur, and he was frequently called upon to testify.

But now, in 1896, at the age of 53, Koch and his young wife were to about to embark on a whole new set of adventures. His Africa years were about to begin.

20

Africa Years: Robert Koch's Research in Tropical Medicine

Well, now, let's shake hands—perhaps for the last time! I'm leaving soon for the tropics and the trip is planned so that I will be in each place at the most dangerous season. Perhaps we'll never see each other again.

—ROBERT KOCH TO CARL SALOMONSEN[1]

I consider it my duty to travel and work where I can use my scientific abilities to the best. At home, there are so many demands on my time, and controversies are so fierce, that it is virtually impossible to get any work done. Out here in Africa, one can find bits of scientific gold lying on the streets. How much have I learned and seen since I first came to Africa!

—ROBERT KOCH[2]

By 1890 the bacterial etiologies of many important infectious diseases had been determined, following the methods and procedures which Koch had developed. In a speech that year, Koch considered the advances which bacteriology had made, and considered where the future would lie. Among other interesting observations made in this speech was the following paragraph:

In many respects, and where we would not have expected it, bacteriology has failed us. We have no knowledge of the causes of diseases like measles, scarlet fever, and small-pox. Of the germs of influenza, whooping-cough, yellow fever, pleuro-pneumonia, and many other undoubtedly infectious diseases, we also know nothing, although skillful work and patient study have not been lacking. I am inclined to think that here the causal agents are not bacteria, but organisms of a far different character. As a partial

237

confirmation of this idea, I can mention the peculiar malaria parasites which have been found in the blood of humans and animals. These parasites belong to the lowest group of animals, the protozoa. We shall know more of these parasites when we can cultivate them in artificial media and study them outside the body. If this can be done, and that they can be cultured there can be no doubt, a whole new field of bacteriology will be opened up.[3]

The point here, and it was a good one, was that bacteria did not explain everything. Most of the diseases Koch listed above are caused by viruses, and malaria by a protozoon. However, looking back from the perspective of the almost 100 years since Koch spoke, we can see that the bacteriological tradition actually hindered, perhaps strongly, the developments of the fields of virology and medical protozoology. Although the requirement in Koch's Postulates that the disease be transferred to experimental animals was a good one, the requirement of artificial cultivation may have hampered research on viruses and pathogenic protozoa.

However, the Kochian orthodoxy spelled out by the Postulates infected Koch's colleagues more than it did Koch himself. Only a few years later, Koch set off on the study of some of these nonbacterial diseases, most of which are tropical in distribution. And when cultivation of these parasites was unsuccessful, Koch was quite willing to continue research on these agents anyway, making strong attempts to control these diseases even without having the etiological agents under culture in the laboratory. Robert Koch embarked on studies in tropical medicine that were to occupy him most of the rest of his life. In 1896 his Africa years were about to begin.

Several critics of Koch, writing in the years after his death, have implied that Koch went off to Africa to avoid the disappointments of the tuberculin discovery and to hide from public criticism of his unorthodox personal life (in those days, a divorce and remarriage were considered shocking). However, there is no evidence that Koch traveled to avoid unpleasantness at home, although it is true that he ceased doing research on tuberculin, and on tuberculosis, after 1896. And it is certainly the case that Koch took full advantage of the travel opportunities which the research on tropical diseases afforded. With his new wife Hedwig (Figure 20.1), he traveled willingly to remote, even dangerous, parts of the world, suffering discomfort and disease in order to pursue his research. And always he continued to send back detailed reports of his work. In fact, there are probably more pages in Koch's collected

Figure 20.1   *Hedwig Koch in Egypt.*

writings on his tropical disease research than on the bacteriological work for which he is most famous.

Koch's name is well known in the annals of tropical diseases. He made many important contributions to the study of malaria, sleeping sickness, and numerous viral diseases of veterinary interest. The only organisms ever named for Koch were protozoa: a trypanosome that was thought to be the cause of Rhodesian red water disease of cattle, *Theileria kochi* (the name was later changed), and an amoeba that was thought to be the cause of simian malaria, *Haemamoeba kochi*. He developed methods for the control of several important cattle diseases in Africa, and his services were widely sought by African governments. In the annals of malaria research, Koch's name is featured along with those of Laveran, Manson, Ross, and Grassi, as one of those most responsible for deciphering this dreaded and important disease. Koch's method for controlling malaria in the tropics was virtually the only satisfactory method until the insecticide DDT became available after World War II. Thus, although today we think of Koch primarily as a bacteriologist, this book would be incomplete if we did not present the broad scope of Koch's tropical years. How did it begin, and what did Koch do?

European colonialism

The end of the 19th century was, of course, the peak of European colonization of Africa and Asia. By this time, almost all of Africa had been occupied by Britain, France, Germany, Belgium, Portugal, and Spain (Figure 20.2).[4] At the outbreak of World War I, Germany controlled German East Africa (later divided between Belgium and Britain), South West Africa (later a South African "Protectorate"), Cameroon, and Togo. Britain controlled British East Africa (later called Kenya), Uganda, Nigeria, the Gold Coast (now called Ghana) and Rhodesia (now called Zimbabwe). South Africa was split into the Cape Colony, which was British, and the South African Republic and the Orange Free State,

Figure 20.2   *Map of Africa just before the start of World War I (1913). The German possessions are shown in heavier outline.*

which were contested but were nominally under the control of the Boers. (The British finally succeeded in uniting all this territory into the Union of South Africa in the Boer War of 1899–1900.) The French occupied most of Northwest Africa and the island of Madagascar, the Portuguese controlled Angola and Mozambique, and the Belgians held the Congo. Some of the greatest agricultural land on earth was (and is) in the highlands of East Africa, and agricultural settlers came in droves.

The settlers, of course, brought their livestock with them from Europe, and it should not be surprising that resistance to native diseases was lacking in these animals. Once a pathogen got started, it could spread through the livestock populations like wildfire. And this is exactly what happened in South Africa in 1896 when rinderpest, a viral disease, broke out. And with rinderpest, came Robert Koch.

South Africa: 1896–1898

Rinderpest, which had been endemic in northeast Africa, made its way across the Zambesi River to South Africa. Despite a barbed-wire fence that was constructed completely across the border, the disease entered the Cape Colony. The seriousness of the disease can be seen from the fact that mortality of greater than 90 percent often occurred once the contagion was introduced into a herd. At the time of Koch's visit, nothing was known about the etiology of rinderpest, although Burdon Sanderson had shown many years before that the disease could be transmitted experimentally by injecting blood from infected cattle.

The British Colonial Office, at the request of the Cape Government, fixed on Robert Koch as the person to bring the rinderpest epidemic under control. At the end of October 1896, Koch received an urgent telegram from South Africa, requesting that he come. Why Koch? At this time, Robert Koch was certainly the most famous medical bacteriologist in the world, and his success in determining the causes of infectious diseases was well known. His previous experience in Egypt and India during the cholera expedition of 1883–84 also meant that he understood tropical conditions, as well as the demands of field research under primitive conditions. But except for anthrax, he had had little experience with veterinary diseases.

And why did Koch accept the invitation? He left behind in Berlin a full institute where a vast amount of important and interesting research was being carried on. But his memories of those pleasant months in

Egypt may have come back to him. Tuberculin had long since ceased to hold his interest. So why not travel? His expenses were to be paid by the South Africans and his salary would remain exactly the same. The German government granted him leave and the direction of the institute was transferred to Ludwig Brieger, one of Koch's most trusted associates.[5]

Koch left Berlin on 1 November 1896 accompanied by his wife Hedwig and Dr. Paul Kohlstock (1861–1901), an army physician who would serve as his assistant. Catching a steamer at Southampton, the party reached Cape Town on December 1. Koch was welcomed enthusiastically by the South African medical community.[6] As one bubbly South African physician remarked: "If the rinderpest has done nothing else than bring men like Professor Koch to the country it has done good work . . ."[7]

At a reception in Cape Town in honor of Koch's arrival, the whole veterinary and medical community turned out. Dr. Te Water, the Colonial Secretary, welcomed Koch in a speech that was reported in a local journal:

> When the Government decided to approach Dr. Koch, he [Dr. Te Water] thought they did the proper thing—(hear, hear)—and the country would, he felt sure, support them in the matter, and agree that they could not have gone elsewhere and fared better. They could only wish Dr. Koch every success in the work he had undertaken—a work which meant the salvation of their pastoral community, and if he failed he was sorry to say that it was his conviction that they would have to give it up as a hopeless task. He might say he felt very sanguine that Professor Koch would before long inform them of some discovery which would make the disaster which was threatening them seem less important and less dangerous. He (Professor Koch) had come to that country with the highest reputation—a reputation, indeed, above that of every scientist he (Dr. Te Walter) knew of. . . . Indeed, he might say that Dr. Koch had come to them with a world-wide reputation—(loud applause)—and he felt confident that with the fact of past experience they might honestly expect that the danger that was now impending might be averted. (Loud applause.) He had great pleasure in proposing "Success to Dr. Koch." (Applause).
>
> Professor Koch's Reply
> Professor Koch (who spoke in German), in the course of his reply, expressed his thanks to Dr. Te Water . . . for the kindly sentiments expressed . . . He knew not whether they would get what they wanted in regard to the discovery of a preventative cure for rinderpest, but he cordially hoped so—(hear, hear)—and if successful not only would he, but all the country, rejoice. (Applause). . . . He much admired the manner in which they were trying to get at the cause of things. In Europe it was

never very easy to persuade people to go to the cause of things and to utilise modern science to that end. In their eagerness to avail themselves of help in that direction he would say unhesitatingly that they were in advance of Europe. (Applause.)[8]

These exuberant speeches epitomize the confident (overconfident!) mood of late 19th century rationalism. Science and medicine seemed, indeed, to offer a solution to any problem.

Doing bacteriology in a primitive field setting is always a challenge. Although Koch had previous experience with primitive conditions when studying cholera in India, the facilities in Calcutta were actually palatial compared to what could be obtained in South Africa. Koch set up his research station near the mining city of Kimberley, in the central part of South Africa (on the border of the Orange Free State). With the enthusiastic assistance of members of the South African veterinarary service, Koch occupied vacant buildings of the De Beers Mining Co. which were in a district free of rinderpest but were near sources of infection. Disease-free cattle were brought in and inoculation experiments begun. Laboratory facilities were set up in a spacious building (Figure 20.3) and Koch immediately attempted to isolate the pathogen.

Previous European work had brought forth claims that the rinderpest pathogen was one or another type of bacterium, and a bacterium had also been isolated by a South African veterinarian. However, despite the fact that the blood of infected animals was highly virulent, Koch could find no evidence of a bacterial pathogen.[9] [The causal agent was shown to be a virus in 1902.[10]] Koch's main work at Kimberley then became an attempt to develop an immunization procedure. The approach used was right out of Louis Pasteur's book. First he attempted, unsuccessfully, to attenuate the virus by passage through animals. Then he tried to reduce virulence by inoculating the virus together with serum from animals which had recovered from the disease. Finally, he achieved partial success by inoculating susceptible cattle with serum from salted cattle plus bile from cattle which had recently died from the disease. However, from a practical point of view, this procedure conferred immunity only for a limited time and on occasion proved fatal to the vaccinated animals. Although one of Koch's German biographers has claimed[11] that Koch's immunization method saved 75 percent of the cattle in South Africa (over two million animals), European workers in other parts of South Africa developed different immunization methods which proved more tractable,[12] so that it is not clear how much of the overall vaccination success was a result of Koch's work.

(a)

(b)

Figure 20.3   *Robert Koch during his first African expedition, South Africa, 1896. (a) Koch in his laboratory in Kimberley. (b) Robert and Hedwig Koch with co-workers.*

However, Koch's brief stay in South Africa ended suddenly. In April, he was off to India on a quite different mission. The total time he had spent in South Africa was about three months. As Robert Koch reported to the British Medical Journal:

I must explain that to break off . . . my rinderpest researches at this moment has not been easy. Although I believe that I have solved the main problem in its essential points, there yet remain a number of secondary questions to answer, and above everything the results obtained at the experimental station have to be tested in actual practice. To have carried this out I should have had to work for some months more . . . but other duties called me, and I could no longer decline the honourable and important task alloted to me of taking part in the work of the expedition sent out by the German Government to investigate the bubonic plague in Bombay.[13]

To India and back to Africa

Shortly after Koch's work on rinderpest began, news came to Europe from Bombay that bubonic plague had broken out. Plague—the black death—was one of the most feared diseases in the world. It had been absent from Europe for many years but was now reported to have reached Naples and Oporto. Three years earlier, the French bacteriologist Alexandre Yersin (1863–1943) had shown the causal agent to be a bacterium. [The organism is now named *Yersinia pestis* in honor of Yersin.] With plague establishing a foothold in two important port cities of Europe, the German government immediately established a commission to go to Bombay and study the disease. The goal of the expedition was to study the mode of transmission of the disease and develop effective control methods. Naturally, Robert Koch was appointed to head this expedition.

However, at the time Koch was appointed to be the head of the German Plague Commission, he was on leave from his German position and was deeply involved in his rinderpest studies in South Africa. Temporarily, another member of the German Commission, Koch's former associate Georg Gaffky, was placed in charge, and the Commission left Berlin for India. Koch himself was ordered to proceed as quickly as possible from South Africa directly to India. Koch thus left his important researches for another lengthy trip to India. He and his wife experienced considerable difficulty getting from South Africa to India; the steamboat lines that normally travelled between the southeast coast of Africa and Bombay had stopped sailing, because of the lack of freight and passengers resulting from the quarantine of Bombay. Thus, Koch reached Bombay by a roundabout route via Mozambique, Zanzibar, and Aden. By the time he reached Bombay, the work of Gaffky's team was well underway and Koch was left to do little more than approve of the control measures which Gaffky had developed (Figure 20.4). A discussion of the results of the German Plague Commission, most of which did not involve Koch, would take us too far afield. A number of immunization studies were undertaken, as well as epidemiological investigations. The main result was a confirmation of Yersin's conclusion that the rat was a reservoir for plague, and an affirmation that the best way to prevent plague was to control rats.[14]

Koch's stay in India lasted only until June and (apparently because of the intense heat at that time of the year) was quite unpleasant.[15] News had reached the commission that plague had also become established in German East Africa, so in early July Koch was telegraphed orders to return to Africa. "This continuation of the journey was too much for my wife," he reported to his friend Libbertz, "and despite my vigorous objections she has gone off with the Plague Commission to Egypt, where she is going to stay with the Kartulis family".[16]

Koch himself arrived at Dar es Salam, the main port for German East Africa, on 12 July 1897. He remained in Dar es Salam for almost a year, studying not only plague but malaria, surra disease (caused by

Figure 20.4   *Robert Koch in India with the German Plague Commission. Georg Gaffky, head of the commission, stands on the left. On the right is Alexandre Yersin, the French discoverer of the plague bacillus (now named Yersinia pestis).*

Trypanosoma Evansi), and Texas fever (a protozoan disease of cattle). The first item of business was a confirmation that the disease in Africa was indeed plague. There was no plague in Dar es Salam, so a lengthy trip to the interior was planned; however, the German government insisted that Koch send his assistant Max Zupitza (1868–1938) rather than go himself. After a strenuous expedition to the interior, Zupitza was able to collect blood samples from native patients and send them back to Dar es Salam. Koch cultured the plague bacillus from these samples and hence confirmed that the disease was indeed plague. However, plague never reached the European settlements along the coast, so that Koch turned to studies of other diseases. He sent back numerous reports, and ultimately published all of these in a book, but there is really little of enduring scientific interest in this work. He returned to Berlin on 19 May 1889, having been away for one and a half years.

Koch's research on malaria

In 1897, the same year that Koch was in Africa and India, Ronald Ross was carrying out his epochal work on malaria in Calcutta, proving that the parasite is transmitted by the mosquito. This work was to have far-reaching consequences for the control of malaria (and was to win Ross the Nobel Prize in 1902, three years before Koch). Laveran had shown in the 1880s that the malaria parasite could be seen microscopically in the blood, and Ross had carried the life cycle to the mosquito. Koch had been interested in malaria for many years, having first studied it in Calcutta during his cholera expedition of 1883–1884. Ross's work on mosquito transmission was with birds rather than humans, and the proof that mosquitoes transmitted the malaria parasite to humans was still lacking.[17] In Europe, malaria was found primarily in Italy, and Koch therefore went to the small town of Grosseto, a few kilometers north of Rome, to begin serious malaria research (Figure 20.5). On this trip he was accompanied by his wife, Richard Pfeiffer, and an assistant, Hermann Kossel. So as to be in Italy during the mosquito and malaria season, the trip was scheduled for the summer.

The party stayed in Italy from 11 August until 2 October 1898, working mostly on the relationship of the mosquito to various forms of malaria in birds and humans. The work mostly confirmed the work of Ross, but Koch had given himself up completely to malaria research, which he carried out for the next year and a half.

Figure 20.5   *Robert and Hedwig Koch about to go off on a search for malaria-bearing mosquitoes, Maremma triste, Italy, 1899.*

Koch returned to Berlin in October 1898 and occupied himself with malaria material which he had brought back from Italy, as well as getting ready for another malaria expedition the following year which was to include not only Italy but also the East Indies and New Guinea.

It must have been when Koch was departing on this trip that the following exchange with Carl Salomonsen occurred. Salomonsen visited Berlin just at the time Koch was getting ready to leave for the tropics and as they were saying goodbye, Koch exclaimed:

> Well, now, let's shake hands—perhaps for the last time! I'm leaving soon for the tropics and the trip is planned so that I will be in each place at the most dangerous season. Perhaps we'll never see each other again.[18]

But he did return.

The Italian part of this trip took place from mid-April to late August of 1899, with residence most of the time at Grossetto. Koch was again accompanied by his wife, as well as two associates, Paul Frosch (1860–1928, later co-discoverer of the foot-and-mouth disease virus) and Heinrich Ollwig (1836–1914). In August, the program was terminated and Koch and his wife traveled south to Naples, where they were to catch a steamer for the Dutch East Indies.[19] The purpose of this trip was to continue the study of malaria in southeast Asia. Koch and his wife were in Batavia in the Dutch East Indies from 21 September until 12 December 1899. From the Dutch East Indies, the Kochs traveled to New Guinea (Figure 20.6). At that time northeast New Guinea was under German control.[20] Koch remained in New Guinea until the summer of 1900. The main accomplishments of this lengthy trip seemed to be statistical and epidemiological studies of malaria, with attempts to relate the incidence of the disease to the mosquito and to develop effective malaria control methods. None of this work dealt with any microbiological or experimental studies. As a result of these studies, Koch developed a method for malaria eradication that involved vigorous application of quinine not only therapeutically but also prophylactically.[21] He wrote to his friend Libbertz that he was very satisfied with the way his work had gone, and that he was confident that he could recommend effective control methods. He then went on to tell Libbertz about his experiences in New Guinea:

> I could fill a book with the things I have experienced on this trip. I would love to tell you everything, but I'll wait until we can have a beer together. New Guinea, at least the part which belongs to Germany, is one of the most beautiful and interesting countries of the tropics. At the moment I

Figure 20.6 *Malaria expedition in New Guinea, 1900. Robert Koch is on the right.*

am resting as a guest of Herr von Bennigsen. From the veranda of his house I have a marvellous view of the blue ocean and a group of volcanic mountains which, if the frequent earthquakes are any indication, might erupt at any moment.[22] All around and below the house are groves of palm trees and in the distance islands with mountains poking up into the blue haze. But, you can't wander unprotected around these palms, as Hedwig has found out. Shortly after we arrived she came down with malaria, and tolerated the quinine so badly that she had to go home in February. The last news I had from her was that she became deathly sick on the trip back but finally reached Naples and went over to Capri to recover. By now she should be back in Berlin, and hopefully is over the fever.[23]

Koch himself returned from New Guinea in June 1900, having thus been away from Berlin for more than a year.

Where did all this malaria work of Koch's lead? Although he did publish one solid paper on the developmental cycle of malaria,[24] most of his basic work simply confirmed that of others. But during the lengthy trip to the Dutch East Indies and New Guinea, Koch formulated a procedure for the control of malaria which he presented in an address to a British medical congress that was held in Eastbourne in 1901.[25] In this address at Eastbourne, Koch began by pointing out the four basic ways by which malaria could be prevented: 1) avoid neighborhoods where malaria is present; 2) exterminate malaria-transmitting mosquitoes; 3) protect dwellings and the body from mosquitoes by use of nets and screens; 4) exterminate the malaria parasite in humans with quinine. He pointed out that although the first three approaches might work in temperate climates, they would not work in the tropics, especially when one was dealing with native populations. He thus concluded that the only effective procedure for the control of malaria in the tropics was by the use of therapeutic and prophylactic quinine treatment.

> In making this proposal, I presuppose two things: firstly, that the malaria-parasites are restricted to man, and, secondly, that we can destroy them, or at least render them harmless, by means of quinine.[26]

He did not, however, advocate universal quinine treatment in the tropics. In Koch's method, anyone suffering from malaria would be treated, and microscopy of blood samples would be used to seek the malaria parasite in those who were well. He developed a simple method for taking blood samples that could even be done by nonphysicians, and the microscopy could also be done by nurses or other nonmedical people. "With a sufficient staff of assistants, a single doctor will be able to rule a pretty large malarial district and rid it of the parasites."

The New Institute of Infectious Diseases

Robert Koch was associated with *four* separate institutes in Berlin. The first was the Kaiserliche Gesundheitsamt (Imperial Health Office) on Luisenstrasse, to which he was allied from 1880 until 1884 (Figure 9.2). The second was the Hygienische Institut on Klosterstrasse, when he was Professor of the University of Berlin (Chapter 17), and where the first tuberculin work was done. The third was the Institute of Infectious Diseases allied with the Charité Hospital, that was established for Koch in 1891 as a result of the excitement over tuberculin (Chapter 19). And the final institute, which is the present-day Robert Koch Institute, is on Föhrerstrasse in northwest Berlin (Figure 20.7). Planning for this last institute was begun in 1896 but construction was not completed until 1900. Most of the planning and construction of this institute was carried out while Koch was in Africa, India, and Southeast Asia.

The initial motivation for the construction of the fourth Institute of Infectious Diseases was that the Charité Hospital needed to expand, and the only place for expansion was in the area occupied by Koch's Institute (the so-called "barracks")[27]. But in addition, an extensive new hospital was being planned, to be called the Rudolf Virchow Hospital (Virchow by this time was deceased), and it was logical to establish the Institute of Infectious Diseases adjacent to this Hospital, since this would provide suitable patients for research. By the time construction of the new institute was underway, Richard Pfeiffer had taken a position as Professor at Königsberg and the scientific department of the Institute was under the direction of Wilhelm Dönitz.[28]

The new building was quite substantial, with a long facade facing the Nord Ufer canal (see Figure 20.7). The three-story building contained laboratories, work rooms, lecture rooms, conference rooms, offices, a room for the culture collection, and a library.[29]

How much influence Robert Koch had on the development and scientific direction of the institute is uncertain, but it could not have been much. He was in frequent correspondence with various colleagues, but none of the published letters seem to provide either scientific or adminstrative direction to the institute director or members. A summary of the activities of the scientific department of the institute at that time,

(a)

(b)

Figure 20.7   (a) *The main building of the Institute for Infectious Diseases on Föhrerstrasse, Berlin. (b) This building still stands (Nordufer 20, Berlin West), although extensively remodeled since World War II. In the left wing is Koch's mausoleum and a small museum. Part (b) photographed in 1987.*

which gives a good overview of what German bacteriology found interesting at the beginning of the 20th century, is given below:

1.   Professor Wassermann. Immunity to hog cholera. Production

and control of sera for erysipelas. Etiology of arthritis. Mechanism of immunity (studies on the side-chain theory of Ehrlich). Studies on the immune agglutination process. Immunity against enzymes.

2. Dr. Moxter. Studies on alexine (an alternate name for complement). Studies on lyssa disease.

3. Dr. Marx. Studies on immunity to cholera and typhoid. Studies on rabies. Tuberculosis studies.

4. Dr. Salzwedel. Disinfection with alcohol.

5. Dr. Neufeld. Immunity to pneumococcus. Pneumococcus as a cause of erysipelas. Assistance with the rabies vaccination clinic.

6. Dr. Schütz. Typhoid fever.

7. Dr. Kolle. Studies on plague.

8. Dr. Ruge. Malaria research in humans and birds.

9. Prof. Proskauer. Dr. Elsner. Dr. Nietner. Studies on water bacteriology.

10. Prof. Beck. Studies on tuberculosis.

11. Prof. Frosch. Malaria research.

In addition to the scientific department, under the direction of Wilhelm Dönitz and Paul Frosch, the following other departments existed: department for highly infectious diseases (cholera, plague, glanders, etc.), under Kolle; serum department, under Wassermann; department for tropical diseases and hygiene, under Schilling; rabies department, under Lentz; and a chemical department, under Lockemann. In addition, the director of the infectious disease department in the Virchow Hospital, Jochmann, was a member of the Institute. A few years later, a zoological laboratory for the study of protozoa was established, as well as a laboratory for the study of smallpox. In addition, the Institute for Experimental Therapy in Steglitz (a district of Berlin) was affiliated with Koch's Institute as a special institute for serum research and serum testing. This Institute, which had previously been under Ehrlich (who now had his own institute in Frankfurt), was under the direction of Wassermann. An indication of the importance over the years of this institute for German medical microbiology can be seen from the bibliography published by Gerber.[30]

Human and bovine tuberculosis

In July 1901 the First British Congress of Tuberculosis was held in London and Robert Koch was awarded the Harben Medal. In his lecture,[31] Koch stunned the Congress with the announcement that human and bovine tuberculosis were two distinct diseases, and that humans did not become infected with the bovine organism. Koch's position on this matter, which went counter to all other opinion of the time, had major public health implications, since if the two diseases were unrelated, then the control of bovine tuberculosis and the pasteurization of milk became less critical issues. Among other things in this paper, Koch said:

> I believe it is no more likely that humans can be infected with tuberculosis from milk, butter, and meat of tuberculous animals than that they can be infected congenitally. I believe, therefore, that control methods for bovine tuberculosis are not necessary.

This was a bombshell! In his first paper on tuberculosis, published in 1882, Koch had declared that bovine and human tuberculosis were due to the same organism. Through the years he had continued to hold this belief. What had caused him to change his mind?

He based his idea on several observations, all of them leading to a false conclusion: 1) children who consumed milk from tuberculous cows only rarely developed pulmonary tuberculosis; 2) the bacterium of bovine tuberculosis differed morphologically, in growth pattern, and in infectivity from that of human tuberculosis; 3) the pathology of bovine tuberculosis is different than that of the human form, being primarily an intestinal disease. Koch therefore felt that the first line of defense against human tuberculosis should be the development of methods to prevent the organism from spreading through the human population via sputum, saliva, droplets, and infected objects.

Koch's former student Behring, who was himself now heavily involved in studies on tuberculosis (among other things, he claimed to have developed a vaccine for tuberculosis, see Chapter 19), took exactly the opposite position. According to Behring, the most important source of infection for childhood consumption (pulmonary tuberculosis) was infected milk. For adults, Behring agreed that a direct infection of the lungs, via atmospherically transmitted droplets, was the rule, but with children, the infection reached the lungs by way of the intestine.

Koch's lecture in London stirred feverish action. Extensive research studies were undertaken in a number of laboratories. A British Commission was established to consider the question, and in Germany a special research group under the Imperial Health Office was established (headed by one of Koch's former students, H. Kossel). Although the British quickly reaffirmed the dangers of bovine tuberculosis, the German commission, after a two-and-a-half-year study, agreed with Koch. The question lingered for many years, and during Koch's visit to America for the International Tuberculosis Congress of 1908, this topic was intensely discussed (see Chapter 21).

Typhoid fever carriers

Koch's contributions to our understanding of cholera have been presented in Chapters 15 and 19. His important discovery of the role of water filtration in the control of cholera (Chapter 19) provided the impetus for the establishment of water filtration plants in many cities. However, although it was the cholera epidemic which provided the insight to the importance of water filtration, it was *typhoid fever* which really was controlled in Europe by water filtration. Whereas cholera occurred in Europe primarily in epidemics, typhoid was endemic.

Our understanding of the epidemiology of typhoid fever is closely linked to the rise of bacteriology as a discipline. The causal agent of typhoid fever had been first seen under the microscope by C.J. Eberth in 1880, and was first cultivated by Georg Gaffky (in Koch's Institute) in 1884. However, it was not until 1902 that Drigalski and Conradi, in Koch's Institute, developed the first culture medium which permitted the selective culture and specific identification of the causal agent of typhoid, at that time known as *Bacillus typhosus*. In 1902 a major epidemic of typhoid fever appeared in Trier, a German city on the Mosel River, and a German Commission was organized under Robert Koch but sent out under Paul Frosch to investigate. Using Drigalski and Conradi's medium, this commission was able to trace the organism through the infected population. The connection of the organism with drinking water and sewage was shown and control measures were introduced to prevent spread from these sources. Although these measures markedly reduced the incidence of typhoid fever, they did not eliminate it in all locations. By careful analysis of the incidence of typhoid fever in a small village in the vicinity of Trier, Koch concluded that the infections were derived not from water or sewage but from other people, people who were themselves perfectly healthy. These people, whom he called "carriers", could be shown to be infected with virulent typhoid bacteria, although the bacteria had no effect on the carriers themselves. This discovery, which was to have far-reaching influence for public health, was announced by Koch in a speech which he gave in Berlin in late 1902. Here is the key passage:

> Our studies have shown that all cases of typhoid of this type have arisen by contact, that is, carried directly from one person to another. There was no trace of a connection to drinking water.[32]

And then came an exceedingly interesting passage. Koch discussed in some detail the efforts that his group had taken to prove that there was no source of infection besides other people. He then referred to his studies in New Guinea, made two years previously:

> The situation here is exactly the same as that which I found during my studies on malaria. In my studies to control malaria in New Guinea, I first showed that there was no other source of infection than people themselves. And this same proof I believe I have shown for typhoid fever here in this small village near Trier.[33]

Making the logical connection between malaria carriers and typhoid carriers was a stroke of genius, and perhaps was a connection that only Koch, with his extensive malaria experience, could have made so readily. Koch's hypothesis regarding typhoid carriers spread rapidly through the western world and was quickly confirmed. It became an important new concept in public health, a concept of extreme importance for certain diseases in which the pathogen may enter into a long-term benign association with a human. The "carrier hypothesis" was one of Koch's last significant bacteriological contributions.

Back to Africa

In late 1902, Robert Koch received an urgent request from the British Colonial Office to come to Rhodesia and study a disease that came to be called *Rhodesian red water*. The appearance of this disease in Rhodesia prompted the High Commissioner for the South African Colonies to prohibit the passage of cattle from Rhodesia into Natal. The background to how Robert Koch, instead of some well-qualified English scientist, came to be invited to study this disease, has been well covered by Dwork.[34]

The details of why Koch was invited, which concern Koch only periph-erally, will not be discussed here. Suffice it to say that Koch was not only famous, he had extensive African experience, and he had made a favorable impression on the South African government during his pre-vious visit. But the financial requirements that Koch set before he would agree to go were stiff:

> Dr. Koch agreed to undertake the investigation on condition that he receive an honorarium of £ 6000 per year (including time for the journey out and home) and a subsistence allowance of £ 3 per day. He requested that two assistants accompany him (F.W. Kleine and F. Neufeld), and that they each be paid £ 1000 per year and £ 2 per day. In addition, he required the payment of all expenses (travelling and equipment), amount-ing to approximately £ 1000. In sum, the estimated expenditure for a one year investigation came to £ 10,000. . . . Cape Colony, originally eager to obtain Koch's services, now felt his terms to be excessive, but the Rhodesian Government was still eager for Koch to come and agreed to pay two-fifths of the costs, leaving £ 6,000 to be divided among the other colonies. . . . Dr. Koch's conditions were officially accepted on December 19, 1902. . . . In light of Koch's modicum of success in 1897 it may appear odd that the colonies were eager to obtain his services once again. It should be remembered, however, that Dr. Koch's method of bile inoc-ulation for rinderpest had been championed by local authorities. . . . More importantly, Robert Koch was, quite simply, the most eminent scientist of that time.[35]

Koch himself viewed the invitation differently. In letters he wrote to friends on Christmas day 1902, just before leaving for Africa, he acted as if he were not very eager to take on the task but was being forced into it. To Flügge he wrote:

> Unfortunately, I have received a commission from the English government that I can't refuse. I must go to South Africa again.[36]

and to Gaffky he wrote:

> I had hoped that I could spend the few years I have left in peaceful contemplation, but it is just not to be. I have received a request from the English government to go to Africa again, this time to Rhodesia, where a new cattle disease has broken out. Of course, there is no thought that I could turn them down, since this is certainly a continuation of the work that I began six years ago in South Africa. My wife will accompany me; we leave on 12 January for Naples, where we'll get a ship.[37]

and to his friend Libbertz he wrote:

> Fate has overtaken me again. I must leave almost immediately for Africa, this time under the auspices of the English government. They want me to study a cattle disease in Rhodesia. Since this is almost certainly a continuation of the work that I began in 1896, I cannot think of turning the job down.[38]

Although in these letters Koch attempted to make his decision to go to Africa scientifically sound, beneath the surface we can hear the sirens of Africa calling him. "Koch would always mention his . . . trips to Africa with a certain tenderness in his voice and declare that the African climate suited him wonderfully. . . . He claimed that the African plateaus had the best climate in the whole world."[39] Africa, in those relatively peaceful colonial days, must have been an exciting delight, especially for one who could travel in style. And Koch travelled in style. With assistants to do the difficult work, a young wife for company, and unlimited black servants, how could one not enjoy? So, when Koch set out for Africa, Frosch was left back in Germany to finish the typhoid study.

The Rhodesian research

Koch published four short reports on his work in Rhodesia, written in English and subsequently translated into German.[40] These reports provide a detailed account of Koch's work, as well as the evolution of his understanding of the disease as the work progressed. Koch set up a base at a camp near the town of Bulawayo, in a heavily infected area. Soon after he was settled, he wrote back to Wilhelm Dönitz in Berlin, the acting Director of the Institute for Infectious Diseases.

> Work is already in full swing. We have set up our laboratory at Hillside Camp, an unused site of the Rhodesian Army. . . . We live in the city and travel out to the Camp twice a day. The trip takes about a half hour. In nice weather it is a very pleasant trip, especially early in the morning. But lately we have had an unpleasant rainy spell, with overcast skies and cold temperatures like we have in Germany. Some days it is fairly warm, but not as warm as Germany in the summer. Evenings it is downright cool, practically winter weather.
>
> We have mosquitoes, but only a few, of the species *Pipiens*. I expected *Anopheles* but haven't seen any yet. Unfortunately, there aren't many pretty butterflies, but lots of ants and locusts.
>
> My wife likes Bulawayo quite a lot, and we are all in good health.[41]

Koch quickly determined that the disease was not, as he had first suspected, Texas Fever, another widespread protozoan disease of cattle. He traced the stages of the protozoal pathogen through the blood and tissues, discovering in the process the schizont phase which subsequently came to be known as "Koch's blue bodies" (from the staining properties of the parasite in the tissues). Having shown that the disease was a new entity, Koch coined the term "African Coast Fever", the name reflecting Koch's conviction that the disease had originated along the East African coast. He developed the hypothesis that the disease had been introduced into the interior by way of a shipment of infected Australian cattle. Evidence was obtained that the parasite was transmitted by ticks, and palliative measures were developed for controlling the disease by dipping cattle to eliminate tick infestations. However, it was difficult to develop a dipping procedure that would completely eliminate tick infestation.

Simultaneously with attempts to control infection, Koch tried to develop a curative serum. However, blood from immune animals proved to be hemolytic, and consequently fatal. Although Koch was able to show that previously infected animals did become immune, he was unable to develop a suitable immunization procedure.

By this time, Koch realized that he would be unable to complete his work within the specified year. He requested, and received, a time extension and an increase in budget, not only to continue his work on the cattle disease, but to pursue studies he had begun on another disease, horse sickness. Koch realized that his absence from home was presenting difficulties, but he was so wrapped up in his work that he was unable to bring himself to leave. In October 1903 he wrote Gaffky:

Dear Gaffky!

My stay in South Africa is going to be a little longer than I had first thought, and I have been remiss in not keeping you informed of my activities. I suspect that you are not happy about my long absence from Berlin and from the Institute. In a certain sense you are of course correct. My absence is certainly of no advantage to the institute. However, although I am unable to carry through with the adminstration of the institute, I reject the charge that I must answer for the damages that might ensue. I had already earlier refused to accept the appointment as Director of the Institute, and (just between you and me) I plan to refuse the position again. But, forgetting about the Institute, I consider it my duty to travel and work where I can use my scientific abilities to the best. At home, there are so many demands on my time, and controversies are so fierce, that it is virtually impossible to get any work done. Out here, one can find bits of scientific gold lying on the streets. How much have I learned and seen since I first came to Africa! And so it still goes. This time I am not involved with rinderpest, but with a group of diseases that are caused by protozoal parasites. These diseases are in such great contrast to the bacterial diseases that we see in Europe: one feels almost as if one is in a new world. . . . If I could, I would stay here for years, trying to get at the root of all these interesting problems. . . . As far as the environment, Bulawayo itself lies north of the 20th parallel, so is really in the tropics, but it is at an altitude of 4500 feet and because of this its climate is quite different from that of the coast. . . . Although it is often cold, the climate overall is quite healthy. My wife, my two assistants, and I find the climate excellent.[42]

On 11 December 1903, Koch celebrated his 60th birthday, far from Berlin and his friends and colleagues. In Berlin a *Festschrift* was published in his honor, with Koch *in absentia*.[43] In a letter to his daughter just after his birthday, Koch commented on the occasion:

I thank you and yours for the birthday greetings on my 60th birthday. This is indeed a special day, as I am now stepping into old age. However, I feel as fresh and vigorous as ever, but I assume that old age will come over me soon. From time to time I have had little indications that everything inside might not be completely all right: heart pains, shortness of breath, but the symptoms don't last. I can no longer climb mountains, but happily, there are no mountains anyway here in the vicinity of Bulawayo. A few weeks ago, though, I was in the District of Victoria, where we were testing my inoculation procedure in an area where there was a large outbreak of cattle disease. In Victoria there were mountains, as well as marvelous granite cliffs. Around one of these large cliffs we saw the remarkable ruins of Zimbabwe, which are said to be from the Biblical city of Ophir. Between our inoculation studies we had time to visit these ruins. But it isn't just the archaeological things. The whole African landscape presents an unforgettable experience: the people, the vegetation, the geology—all together carry a truly African character. And we even have lions! When I get back to Berlin, I will tell you everything about it.

Just recently I returned from a trip to Pretoria and Bloemfontain to attend a rinderpest conference. And imagine this: I had to travel five days by train to get there, ten days round trip. Even that is African in character, because no one here worries about distances. But let's not write anymore about my trips. It is almost Christmas time and I want you to give to each of your sons a hundred Marks as a Christmas present from me. I'll pay you back when I return.

I hope I will have my work wrapped up around the beginning of March, and then I will leave. First several weeks in Dar es Salam and then back in Berlin by May. In this way, I'll miss the German winter, which I especially detest when coming from the sunny southland.

With best wishes to Eduard and the children,

Your loving father.[44]

Koch continued his work until the end of February 1904, at which time he prepared to leave. In his final report, he described an inoculation procedure in which herds were treated with 5 cc of defibrinated blood once every two weeks for four to five months. Whereas uninoculated cattle had a mortality rate of 90 percent, most of the treated cattle survived.

Koch left Bulawayo in March 1904, believing that he had found a solution to the problem of Rhodesian red water, but subsequent work by the South African veterinarians Stewart Stockman and Arnold Theiler determined that Koch's inoculation method was impractical and perhaps not even effective. These workers recommended a massive, highly organized control program which involved fencing, spraying the range to reduce ticks, and dipping the cattle to destroy the tick vectors. The main aim of this program was to prevent the disease from developing rather than to develop a breed of South African immune cattle carriers. Part of the motivation for this approach was to keep South African cattle from being isolated from world markets (which would have been the case if they had been immune carriers of the parasite). One of the veterinarians who had worked with Koch, C.E. Gray of Rhodesia, made the following statement:

> I regret to say that while Dr. Koch's prediction that herds infected at the time when inoculation was begun would not have benefitted thereby, has been fulfilled, neither have those herds been protected which were clean at the time we started, nor has the percentage of mortality in such herds been diminished when the disease disappeared. . . . Repeated inoculations with the blood of recovered animals, as recommended by Dr. Koch, has failed to afford the protection against African Coast Fever which we all hoped would follow such treatment. It does no harm, it does not communicate the disease, but it fails to protect, therefore I cannot conscientiously recommend the public to depend upon such a method of inoculation.[45]

Thus, as far as the South Africans and Rhodesians were concerned, Koch's work was a failure, although his work significantly advanced understanding of this complex parasitic disease. Even today, the problem has not been completely solved.[46]

In Retirement, and Back to Africa

On 1 October 1904 Robert Koch officially retired as Director of the Institute for Infectious Diseases. As we have noted, he had hardly been "directing" the institute since 1896, and he had desired to retire long before. But now Robert Koch was appointed an Honorary member of the Institute and his duties were officially ended. Even after retirement, however, he maintained a laboratory at the Institute and continued some work, but for all practical purposes he was finished. Georg Gaffky, Koch's long-time associate and friend, was brought back from Giessen to be Koch's successor. Thus, although Pfeiffer and Dönitz had been "acting" directors of the institute, Gaffky was officially the second director of the Institute for Infectious Diseases.

Koch did not remain long in Berlin. By the end of 1904 he was back in Africa again, this time to study not only African Coastal Fever, but recurrent fever and African sleeping sickness. This time he was working not in British territory but in German East Africa, under the auspices of the German government. He set up his headquarters first in Dar es Salam, but later moved to Amani in the highlands because the climate was more favorable than that of the coast. He studied the transmission of these diseases, with special attention to the role of arthropod vectors. Recurrent (relapsing) fever is caused by a spirochete and had been one of the first infectious diseases for which a bacterial cause was indicated. As early as 1868, Otto Obermeier (1843–1873) had observed the characteristic spirochetes in blood and in 1878 Koch had published early photomicrographs of the spirochete, now called *Spirochaeta obermeieri*.

African sleeping sickness was first shown to be caused by a trypanosome, *Trypanosoma gambiense* in 1901 by R.M. Forde, and to be transmitted by the tsetse fly by David Bruce (1855–1931) in 1903. The work on African sleeping sickness followed in the tradition of Ronald Ross (1857–1932) and Patrick Manson (1844–1922) on malaria (in India and Africa) and Theobald Smith (1859–1934) on Texas cattle fever (in the United States). Smith had shown in 1893 that the cause of Texas fever was a protozoon and that the parasite was transmitted by ticks. It was natural for Koch, who had worked so extensively on malaria in the late 1890s, to find interest in and pursue studies on African sleeping sickness, especially since this disease affected the colonization of Africa by Europeans.

Koch's own work on sleeping sickness in 1905 was directed primarily at attempts to develop immunization or prophylactic measures. Although Bruce had made the connection between the tsetse fly and sleeping sickness, he thought the role of the fly was purely mechanical. Koch showed that the trypanosome went through cyclic changes in the fly, but he also described a purely imaginary sexual cycle.[47]

At the end of October 1905, Koch returned to Berlin to prepare for a very important occasion: he was to receive the Nobel Prize in Medicine!

The Nobel Prize

The first Nobel Prize for Medicine was awarded in 1901 to Emil von Behring for his work on antitoxin therapy. As we have seen (Chapter 19), Behring was a student and associate of Koch who (together with Kitasato) discovered antitoxins. Although the initial practical development of diphtheria antitoxin was carried out in Koch's institute, Behring soon went off on his own, eventually becoming a professor at the University of Marburg. Together with Paul Ehrlich (who did not receive the credit he deserved—see Chapter 19), Behring developed diphtheria antitoxin into a practical and very important therapeutic measure. His work certainly had far-reaching significance and it was understandable that he might receive the Nobel Prize. Koch, as we have seen, had made a number of enemies, and his reputation had also been seriously tarnished by the tuberculin scandal (see Chapter 18), so it is understandable that he might have been passed over for the first prize. But any dispassionate observer could readily agree that Koch's work was at the very *basis* of modern medicine and public health, and was definitely "prize-worthy". The fact that Behring received the prize before Koch may have reflected the philosophy of the prize Committee that practical rather than fundamental achievements should be recognized.

In 1902, the Nobel Prize for Medicine was awarded to Ronald Ross for his very important work on the role of the mosquito in the transmission of malaria. As we have seen earlier in this chapter, Koch was also involved in the malaria work. Although his studies were definitely derivative of Ross, they were very important for Ross' whole case. In 1903 the Nobel Prize was awarded to Niels Ryberg Finsen (1860–1904) for his work on the use of phototherapy (light rays) for the treatment of diseases of the skin. In 1904 the prize was awarded to Ivan Petrovich Pavloff (1849–1936) for his fundamental and very important work on conditioned reflexes.

But where was Robert Koch? In 1905 Fritz Schaudinn (1871–1906) was nominated for the prize. Schaudinn had published important papers on protozoology (some of which later proved to be erroneous), but in the spring of 1905, working with Erich Hoffman (1868–1959), Schaudinn announced the discovery of *Spirochaeta pallida* (later called *Treponema pallidum*) as the causal agent of syphilis. At the end of April 1905, Elie Metchnikoff, at that time Director of a laboratory at the Pasteur Institute, wrote a letter to the Nobel Prize Committee regarding Schaudinn's nomination:

> Although I think highly of the work of Schaudinn, it is impossible for me to support him as a candidate for the Nobel Prize. I have nominated Koch for the prize for years and as long as Koch has not received the Prize, I can on principal support no other candidate. It is my opinion that Robert Koch's service to medicine has far surpassed that of all other possible candidates.[48]

The Nobel Prize Committee finally came to its senses. On 12 December 1905, one day after his birthday, Koch received the prize (at that time worth 150,000 German Marks). As the *Deutsche Medizinische Wochenschrift* remarked:

"No one worthier can be imagined."

Koch's address in Stockholm was entitled: *On the present status of tuberculosis*. It was a review of recent work, none of which Koch himself had personally carried out. He wrote his old friend Libbertz of his impressions of the trip to Stockholm:

> Many thanks for your wishes on my birthday and on the Nobel Prize. I received the latter after a rather strenuous trip to Stockholm. You are well aware that I am no champion of fancy ceremonies, since one must eat when one is not hungry and drink when one is not thirsty. But my wife had a great time, revelling in all that Stockholm had to offer.[49]

Back to Africa again

Shortly after returning from Stockholm, Koch began to make preparations for another trip to Africa, to continue his work on African sleeping sickness. The trip was supported by the Imperial Health Office and the funds came from the German Reich (the Reichstag itself had to appropriate the funds). This was to be a major expedition, and Koch spent most of the winter rounding up the funds and arousing interest by lecturing on his earlier work. Koch's expedition left in mid-April for

Figure 20.8   *The German Sleeping Sickness Expedition, 1905–1907. From the left: Kudicke, Kleine, Koch, Beck, Panse, Sacher.*

Figure 20.9   *Robert Koch on the Sese Islands, Lake Victoria, during the Sleeping Sickness Expedition of 1906.*

East Africa. He was accompanied by his wife, as well as Prof. R. Kudicke, Dr. F.K. Kleine, Prof. Max Beck, and Dr. O. Panse (Figure 20.8). On part of the trip, Koch's old friend from Egypt, Alexander Kartulis, accompanied him. Also, his friend Libbertz tagged along as an unpaid volunteer.

This lengthy expedition extended from May 1906 until November 1907. Hedwig Koch succumbed to malaria and had to be sent home early, but Koch himself was away from Berlin for 18 months. He kept detailed diaries which were subsequently published, and there were many photographs taken (Figure 20.9), so that this was one of the best documented of Koch's expeditions.

Research stations were set up once again at Amani (near Tanga in the coastal zone of German East Africa), Muanza (a large German station on Lake Victoria), and the Sese Islands in Lake Victoria, near Entebbe in British East Africa (now Uganda). The work was mainly concerned with the epidemiology of African sleeping sickness, a study of the tsetse fly, and the mechanisms of transmission of the disease.

Koch returned to Berlin in November 1907 and occupied himself with compiling his detailed notes. He did not know it, but he had taken his last trip to Africa. However, in the two years left to his life, he would travel almost completely around the world and see two countries he had longed to see, Japan and the United States.

21

The World Tour: Koch in America and Japan

If I think today of all of the praise which you have heaped upon me, I must, of course, immediately ask myself if I deserve it. Am I really entitled to such homage? I guess that I can, with a clear conscience, accept much of the praise you have bestowed upon me. But I have really done nothing else than what you yourselves are doing every day. All I have done is work hard and fulfill my duty and obligation. If my efforts have led to greater success than usual, this is due, I believe, to the fact that during my wanderings in the field of medicine, I have strayed onto paths where the gold was still lying by the wayside. It takes a little luck to be able to distinguish gold from dross, but that is all.

—ROBERT KOCH[1]

While Robert Koch was far off in Africa studying sleeping sickness, at home on 24 March 1907 a celebration was held in honor of the 25th anniversary of the announcement of Koch's discovery of the tubercle bacillus. This 25-year celebration[2] inspired the establishment of the Robert Koch Foundation for the Conquest of Tuberculosis. A commission of noted German health officers and doctors was established, and a goal was set to collect a large sum of money. An international fund-raising drive was initiated, the prospectus noting how appropriate it was that the fund was to be named in honor of Robert Koch, "one of the greatest researchers of all times."[3] From darkest Africa, Robert Koch himself wrote in support of the Foundation.

Many wealthy German citizens made significant financial contributions, including His Majesty the Emperor (100,000 Marks), but the largest donation of all was that made by the American philanthropist Andrew Carnegie, who in 1908 donated 500,000 Marks. This donation was made in anticipation of Robert Koch's visit to America and the discussion of Carnegie's gift was one of the highlights of a dinner held in Koch's honor in New York (see later).

In retirement, Robert Koch planned a lengthy trip to the United States and Japan. His goal in the United States was primarily to visit his brothers who had left Germany years earlier and settled in the Midwest, and then continue across the country to San Francisco, where he could get a steamer to Japan. After Japan, his original intention was to return home to Europe by way of China, thus having travelled completely around the world. This latter part of the trip was thwarted because he was ordered to return to the United States to participate in the International Tuberculosis Congress in Washington, as an official delegate of the German government. As we will see later in this chapter, this second visit to the United States led to serious controversy.

Koch travelled to America by way of London, where on 9 March 1908 he attended as an official German delegate an international conference on African sleeping sickness, reporting on some of the results of his large African expedition. Then on 29 March, Koch and his wife departed from London for the United States, arriving in New York on 8 April 1908, where they were greeted enthusiastically.

Koch's First Visit to the United States

> Dr. Robert Koch, the famous bacteriologist, arrived here yesterday on the North German Lloyd liner Kronprinzessin Cecilie, accompanied by his wife, on a tour of the world. As the liner came up the bay, he stood on the boat deck taking in a view of the harbor, hemmed in by reporters and photographers.
>
> "Let me see the harbor!" he exclaimed, "and then I will talk."
>
> In appearance Dr. Koch is a typical German professor, with gray hair and beard. A pair of bright blue eyes twinkle through gold-rimmed spectacles. Mrs. Koch is an attractive blonde, much younger than her scientific husband [Figure 21.1]. She takes a deep interest in his work, and has accompanied him on many of his long, hazardous journeys in search of the elusive bacillus.[4]

Koch was met at the Quarantine Station by a committee from the German Medical Society of New York. Mrs. Koch was presented with a bouquet of roses and Koch was asked to make a statement for the New York papers:

> "I expect to stay in New York until April 11 or 12," said the distinguished visitor in perfect English. "From here I shall go to Niagara Falls, and from there to Chicago to see a brother who lives in that city. I shall go

Figure 21.1   *Robert and Hedwig Koch in New York at the beginning of their United States visit.*

thence to St. Louis, and on to San Francisco as quickly as possible, because I wish to get to Japan. Most of my time will be spent in that country, as it has the most interest for me from a scientific point of view.

"I have been to Africa five times in the course of my work, and hope I shall not have to go there again. That is not certain, however, because the fourth time I went I said the same thing, but in a few months after my return to Berlin my plans were all made to start again."

Koch was asked what he thought of a vivisection bill that was before the New York Legislature.

"You can do what you like in America about vivisection, but my advice is to do as the people do in Germany and let it alone."

Upon being asked his opinion about Metchnikoff's theory for prolonging life on a sour-milk diet he answered:

"I know him personally and esteem him very highly, but I never express an opinion upon subjects with which I have had no experience."

Mrs. Koch came to the rescue of her husband, and told the reporters some of their experiences when she accompanied him to East Africa in search of the germ of the "sleeping sickness" that has caused so many deaths among whites and blacks in that region.

"We are taking this long vacation," said Mrs. Koch in excellent English, "because my husband had such a strenuous time during the last eighteen months he spent in Africa on an island in Lake Victoria.

"I spent nine months of that time with him on the island, but I was taken ill and had to go down to the coast to recuperate. He went there to study the sleeping sickness, which had created such havoc in the German colonies and is now ravaging the British colonies. . . ."

From the pier Dr. and Mrs. Koch were driven to the Plaza, where they will stay while in New York. Besides a visit to the Museum of Natural History, Dr. Koch will inspect many of the hospitals of the city. . . .

Dr. Koch is in his sixty-fifth year, but is still in the vigor of life and very active. He revels in work, and during his recent trip to Africa he frequently spent twelve to fifteen hours a day in his improvised laboratory.

Mrs. Koch admitted she was not a scientist, but naively added, "What woman could help taking an interest in it if she was married to a man like that?"[4]

A Festive Dinner

Robert Koch received a rousing welcome at a banquet arranged in his honor by the German Medical Society of the City of New York. The banquet was held at the Waldorf-Astoria Hotel on Saturday evening, April 11, 1908, with about 450 in attendance (Figure 21.2). On the right of the guest of honor sat Andrew Carnegie. On the Banquet Committee were such luminaries of United States medicine as Simon Flexner, head of the Rockefeller Institute for Medical Research, William Henry Welch, Professor of Pathology at Johns Hopkins University, Theodore Darlington, Commissioner of Health for the City of New York, Theobald Smith, Professor of Pathology at Harvard University, Edward L. Trudeau, head of the Tuberculosis Sanitorium at Saranac Lake, General George M. Sternberg of the U.S. Army Institute of Pathology, and Hermann M. Biggs, head of the Health Department of the City of New York. The menu was French.[5]

The principal addresses at the banquet were given by William Henry Welch, Professor Abraham Jacobi, Andrew Carnegie, and Koch himself. In leading off the speeches, Welch recalled his first encounter with Koch many years before at Breslau, before Koch had achieved fame (see Chapter 6).

Dinner

in honor of His Excellency
Professor Robert Koch,
under the auspices of the
German Medical Society
of the City of New York.

Saturday, April 11th
1908

The Waldorf-Astoria.

Figure 21.2   *Announcement of the festive banquet in New York.*

I remember so well his visit at that time to Professor Cohnheim's laboratory; how they passed into a little room, and after Professor Koch had departed Dr. Cohnheim told us of their talk in the laboratory; that he was a great man, who had done a sensational piece of work to be heard of throughout the medical world.[6]

Welch also recounted his attendance at Koch's bacteriology course at Berlin, in the Hygienic Institute:

Prudden and I appeared there in the first semester of 1885 [see Chapter 17 for Prudden's account of this course]. . . . In speaking with him last night, we were rather surprised to hear him say that he didn't consider himself a good teacher. That was not our experience, nor the experience of his pupils at that time. If good teaching is to inspire enthusiasm on the part of the pupils, if it is to impart that real living knowledge which is a part of your flesh and blood, if that is good teaching, then Professor Koch is not only a great investigator, but he is also an accomplished teacher.[7]

Welch described Koch's career in detail, commenting especially on the enlightened attitude of the German government in recognizing Koch's genius and giving him his first position in Berlin.

> Is there a period in the history of medicine where such discoveries issued from one man, and from those working under his supervision; or, in the whole history of medicine, is there a like period where such discoveries are found as the laying of the foundations of modern bacteriology, forging the instruments with which we work to-day, exploring these newly discovered fields, demonstrating the specific micro-organism of tuberculosis, the greatest discovery in the whole field of bacteriology, the entire field of which has not yet been fully reached?[8]

Welch then went on to show how deeply Koch's work affected the whole field of medicine and public health, ending on this important note:

> Medical science can make an appeal to-day for aid and support which was impossible before the work of Robert Koch. We can appeal to Mr. Carnegie and Mr. Rockefeller as never before, and the appeal has been answered. And I should like to express, in behalf of the whole medical profession of this country, and especially of all of those present, to Mr. Carnegie, who is, fortunately, with us to-night, our great obligations and especial thanks for the contributions which he has made to that foundation which bears the name of our guest, and in which all of us have been so deeply interested.
>
> We pay our tribute . . . to one of the greatest benefactors of his time; we pay our tribute to the greatest ornament of our profession, and to one of the most important contributors to science, and I raise my glass, and I ask you to drink to German medicine and German science, embodied so worthily, and with so much lustre, in our guest, Robert Koch.[9]

After William Henry Welch's speech, Andrew Carnegie himself rose. After regretting that he could not speak German (he said, in effect: "How many millions of dollars would I give if I could be enabled to instantaneously speak German"), Carnegie lauded scientists as the real heroes of society:

> Gentlemen, we have made heroes hitherto, until recent date, of none except those who had killed their fellows. . . . Now, gentlemen, that day is passing. . . . Imagine the change that has occurred in the space of our lives. Fifty years past, suppose the French people had been polled as to who was the greatest Frenchman, the greatest hero that had ever lived. You all know, Napoleon. Last year more than two million adults were polled in France by one of the newspapers, and what do you think was

the result? Napoleon was the seventh on the list! Whom do you suppose was first? Pasteur. And who was second? An author, Hugo. . . .

Who do you suppose are [my] heroes? The modern hero is he who has served or saved most men, women and children—Harvey, Pasteur, Jenner . . . and last, but not least, Koch.[10]

Finally, Robert Koch himself rose to speak. Despite his good English, he spoke German (it was, after all, the German Medical Society of New York):

If I think today of all of the praise which you have heaped upon me, I must, of course, immediately ask myself if I deserve it. Am I really entitled to such homage? I guess that I can, with a clear conscience, accept much of the praise you have bestowed upon me. But I have really done nothing else than what you yourselves are doing every day. All I have done is work hard and fulfill my duty and obligation. If my efforts have led to greater success than usual, this is due, I believe, to the fact that during my wanderings in the field of medicine, I have strayed onto paths where the gold was still lying by the wayside. It takes a little luck to be able to distinguish gold from dross, but that is all.

We have achieved all we could in our fight against tuberculosis. We have come to a point where we could hardly hope for more success. The idea of building sanatoria will not accomplish much; such sanatoria will only benefit certain localities. We must make new researches. Such researches will become possible in the Robert Koch Institute for Tuberculosis in Berlin, a foundation which Andrew Carnegie has so munificently endowed. In this institution investigations will be made which will open new fields, new theories, new modes, and possibilities of fighting the old enemy, tuberculosis. It will be an international affair, benefiting all mankind. And we have to thank Mr. Carnegie for placing this institution on a sound financial basis. *Herr Carnegie lebe hoch!*[11]

Those were the days! Before the world started collapsing with World War I. When science could do anything. And when private philanthropy seemed unlimited.

A visit to the New York Department of Health

Robert Koch was naturally interested in the manner of operation of the New York City Department of Health. Not only was this the best and most modern in the United States, but Hermann Biggs, its director, and William H. Park, the Director of the Research Laboratories, were ardent followers of Koch and his work. Biggs had established extensive tuberculosis control methods and Park was the first to demonstrate the importance of carriers for diphtheria. Park also played a major role in introducing the use of diphtheria antitoxin into the United States. Koch's visit has been reported in Winslow's biography of Biggs:

> Koch showed great interest in several large statistical tables showing New York City's figures for tuberculosis. Koch pointed to the mortality curve, which, he contended was more and more assuming the characteristics of a parabola. He was quite pessimistic and remarked that unless some new effective weapons were devised in the fight against tuberculosis, we should soon arrive at the point where the tuberculosis mortality would remain practically stationary.
>
> "And Biggs," Koch spoke (in German). "We are all waiting for you. What are you going to do next?".
>
> Biggs also expressed pessimism: "At the present time, most of our sanitorium work is wasted, for usually we are compelled to return the healed patients to the same environment from where they have come and which has been largely responsible for their physical breakdown."[12]

In the further course of his visit, Koch became quite impressed with the ways in which the American health authorities were giving bacteriological assistance to the practicing physician. For instance, in New York City, a physician could leave a throat culture at a corner drug store at 5:00 P.M., it would be picked up by the health department, and he would receive the report by telephone before 10:00 A.M. the next morning.

> You will agree, my dear Biggs, that most of these bacteriological and serological discoveries have come from Germany. For my part I must admit with shame that we in Germany are years and years behind you in their practical application. You have done marvelous work![13]

Travels through the Midwest

Robert Koch had told the New York Times reporter, perhaps tactlessly, that his main scientific interest was Japan rather than the United States. Indeed, he spent very little time in New York, and visited no other cities on the east coast. His main interest in visiting the United States was social rather than scientific. He had two brothers in the Midwest, one a grocer in St. Louis, the other a farmer who lived near Keystone, Iowa. A brief report of Koch's visit, including a rare photograph (Figure 21.3), can be found in an Iowa medical journal.[14] The story of the visit was

Figure 21.3 *Robert Koch in Iowa. From left to right: Adolph Koch, Hedwig Koch, Henry Koch, Mrs. Adolph Koch (seated), Robert Koch.*

told by Doctor O.W. King, at that time a practitioner in Keystone, Iowa. The visit to such a small town by such a famous person was of course of great interest, and Dr. King was very interested in meeting the discoverer of the tubercle bacillus. Also, since Dr. King owned the only automobile in town, he was asked to come on Sunday to Adolph Koch's farm, bringing the local barber, who had a good camera, to take the family picture. Koch at first objected strongly to the taking of the photograph, but finally consented after exacting a promise that only a limited number of copies should be made. Koch also avoided all contact with newspaper reporters, and frowned on any publicity, so that very few knew of his visit.

On to Japan

The Kochs travelled by train from St. Louis to San Francisco, and left San Francisco by steamer for Honolulu on 9 May 1908. They spent two weeks in Hawaii before travelling on to Japan. From Honolulu, Koch wrote Gaffky:

> Here we are in the middle of the Pacific Ocean, at Honolulu in the Sandwich Islands. The trip has been exhausting and we are taking several weeks to recuperate before moving on. The trip across the Atlantic Ocean was awful, the weather varying from heavy to very heavy seas. And then the long trip across America, with many night trains. We were beset wherever we turned by rude reporters. Every time we turned around the temperature was different, so we had to keep changing clothes and packing and unpacking our suitcases. Finally, we ended up in San Francisco in the midst of a mass of humanity which had come to celebrate the return of the U.S. fleet, which meant, in America, celebrating as noisily as possible.
>
> But now we are happy again, sitting here in peace and quiet, able again to lead a contemplative life. Honolulu reminds me of Colombo, with its surroundings and its completely tropical vegetation with coconut palms, and yet, it is not too warm. Our hotel lies right on the beach and swimming is pleasant and comfortable.
>
> We are leaving here 1 June and it will take 11 days to reach Yokohama, without a stop on the way. But that is the longest sea voyage which we will have, since on the way back the rest of the trip will be in shorter stages.[15] I'll write you again from Japan.[16]

Robert and Hedwig Koch were welcomed in Japan with typical Japanese hospitality by Koch's former student and colleague, Shibasaburo Kitasato (Figure 21.4). Kitasato was director of the Institute for Infectious Diseases in Tokyo,[17] and Koch viewed with interest all of the research under way. They were housed in a small European-style hotel at Kamakura, a small seaside resort south of Tokyo and Yokahama. And then, after some days observing Japanese research, they took a two-month tour of Japan. On this tour they were well hosted since Kitasato and two of his assistants accompanied them on the whole trip and in each city the Koch's were feted by the local medical establishment. Koch wrote to his daughter: "My long-held wish to see Japan has finally been granted. We have been travelling in Wonderland for almost two months and every day there has been something new, beautiful, and interesting to see."[18]

As noted, Koch had intended to return to Germany by way of China and India, but his plans had to be altered because he received an urgent request from the German government for him to return to the United States as an official delegate to the International Tuberculosis Congress, which was to be held in September-October in Washington. Koch was

Figure 21.4   *Robert Koch in Japan. (a) Robert and Hedwig Koch among the workers at the Kitasato Institute of Infectious Diseases. Shibasaburo Kitasato is sitting next to Robert Koch. (b) Hedwig and Robert Koch with Kitasato and his family.*

quite displeased, but felt himself bound to agree. He wrote to his daughter:

> I had planned to visit China and not return home until early next year, but it was not to be. I just received a telegram requesting me to return to Washington and unfortunately, I must go back by the same way that

I came. So my plan to go around the world has come to naught and I will have to make another stab at this later. However, despite this disappointment, the trip has been very valuable. I have seen Japan, which is, to me, one of the most beautiful and interesting countries in the world. And I had a chance to see my brothers in America. Now I will be back in Berlin around the end of October.[19]

It might seem strange that Koch, firmly retired from his official position, would feel obliged to accede to the wishes of the German government. As he explained it to one of his colleagues:

It was hard, indeed, to decide to accept the Ministry of Culture's request that I be an official delegate at Washington. But what should I do when the Herr Minister telegraphs that he would value very highly my participation in the Congress? Such a bother. When we came to Japan it was still in the rainy season, and then came the hot summer weather, which will last through September. And after that, we could have expected a period of really nice weather, when further travel through Japan would have been at its best. So I had held off seeing many interesting things, waiting for the fall. And now suddenly my plans are shot and I must return home by the same way that I came, just to please the Americans and add a little weight to their Congress. Indeed, I don't really know what I am suppposed to do in Washington. . . . I have had nothing to do with the organization of the Congress, and I don't even know the program. They wrote me that the Congress won't even start in Washington, but in New York, and in addition there will be meetings in Philadelphia. The whole thing seems a bit of a mess.[20]

The International Tuberculosis Congress

Why was Robert Koch so urgently required at the International Tuberculosis Congress (Figure 21.5)? The main issue, and it was an extremely important one, was Koch's position that the organism which caused bovine tuberculosis was of little danger to humans. As we have noted (see Chapter 20), the public health implications of Koch's position were major, and there was good reason to believe that Koch was wrong. But Koch's opinion was held in such high regard that it could well counterbalance the massive weight of solid scientific evidence.

As discussed by Maulitz;[21] the issue can be viewed as dealing with the mutability and intertransmissibility of the human and bovine strains of tubercle bacillus. In his first work, Koch had insisted on the similarity or identity of these two organisms, but then he had changed his opinion

Figure 21.5   *Robert Koch in Washington, 1908, for the International Tuberculosis Congress. The photograph was taken in front of the German Exhibition (posters in the background). To the right of Koch is William Henry Welch. Second from the left in the back row is Theobald Smith. Hermann Biggs is fourth from the left in the back row.*

in 1901 when he spoke at the International Tuberculosis Congress in London. Even though a British commission had concluded after the London meeting that Koch was wrong, his opinion held much weight, and was the accepted fact in Germany. The United States followed carefully the German lead in most matters bacteriological, and was hence strongly on Koch's side. But one important American, Theobald Smith (1859–1924), had insisted that although the two bacterial forms were distinct, they were both infectious for humans. He took the position that the tubercle bacillus exhibited a degree of polymorphism, allowing it to adapt to the host in which it had been inoculated. Smith, who had begun his career at the U.S. Department of Agriculture, was cognizant of the potential hazard to humans from dairy cattle and promoted a strong movement to purge cattle herds of tuberculosis.[22] Koch remained skeptical, and it was apparently because of this that his presence in Washington was requested.

In the main session of the Congress, Koch gave a paper entitled: *The Relations of Human and Bovine Tuberculosis*[23] in which he reiterated all his old arguments. After reviewing the earlier work, Koch went on:

Many of my opponents have made strenuous efforts to prove that tubercle bacilli in man and those in cattle cannot be of different species, and they imagine that they can thus refute my contentions. I have never held that we are dealing with two distinct species, but have only stated that they differ from each other in certain characteristics which are of the greatest importance in combating tuberculosis. Whether these differences justify one in speaking of special varieties, or even species, is, from my point of view, quite irrelevant. I am concerned only with the practical significance of the differences between human and bovine tuberculosis. . . .

For combating tuberculosis, it is absolutely without significance what changes tubercle bacilli will undergo after being passed through a series of animals or during cultivation under some artificial conditions. Those men who consume milk and butter do not hold back and make cultural or animal experiments; they eat them in the fresh, unchanged condition. In my opinion, therefore, we are concerned here with the properties of the fresh and unchanged tubercle bacillus only.[24]

This rather remarkable statement is perhaps understandable when we recall that Koch's whole career was built on animal experimentation and on the establishment of infections in experimental animals. Indeed, the very identity of a bacterial pathogen required that it *not* change when passed through an animal. However, the point here, as Koch's opponents recognized, was not an abstract theoretical one, but an intensely practical one. The controversy, indeed, boiled down to the fact that Koch was concerned with classic tuberculosis, pulmonary tuberculosis, whereas his opponents were concerned with serious infection of people with the tubercle bacillus, whether or not the symptoms were pulmonary. In the case of transmission of the tubercle bacillus from cattle via milk, which occurred most commonly in children (since they were the primary milk consumers), the infection was more likely to be intestinal than pulmonary. As one of the British commenters on Koch's paper said:

Prolonged and careful experimental research has demonstrated beyond doubt that Koch was right when he affirmed that human and bovine tuberculosis are not identical, but it has been demonstrated that the great scientist was in error when he said that bovine tuberculosis could not be communicated to the human. A large amount of tuberculosis affecting the human body is of bovine origin, and consequently a different line of treatment is required, as compared with the treatment of pulmonary tuberculosis.[25]

In response, Koch answered, rather irrelevantly:

Of all human beings who succumb to tuberculosis, eleven-twelfths die of consumption, or pulmonary tuberculosis, and only one-twelfth of other forms of the disease. One would have expected, therefore, that those investigators who are interested in establishing the relations between human and bovine tuberculosis would have searched for bacilli of the bovine type preferably in cases of pulmonary tuberculosis. This, however, has not been the case. Evidently animated by the desire to bring together as many cases as possible of bovine tuberculosis in man, they have investigated particularly cases of glandular and intestinal tuberculosis, and have neglected the much more important pulmonary tuberculosis. . . . The gist of it is that up to date in no case of pulmonary tuberculosis has the tubercle bacillus of the bovine type been definitely established. . . . On account of the great importance of this question, I intend to undertake, as soon as feasible, experiments along this line on a broad scale. At the same time I wish to make the plea to other tuberculosis workers that as many cases as possible be examined and to join with me vigorously in this task. But I wish to lay stress on the fact that the conditions laid down by me for the carrying out of these investigations must be followed.[26]

In response, an American worker at an important Agricultural College said: "On a correct solution of this question depends the health and lives of many children. I would consider it an extreme misfortune to this and every other country if any impression should go out from the congress that even a small proportion of deaths due to the bovine bacillus is a negligible quantity."[27]

The closed meeting

The open meeting of the Congress at which Koch spoke led to no general agreement. The American tuberculosis community held Koch's views in highest esteem and hence arranged a special conference to discuss the matter further. This special closed-session conference, held at the Willard Hotel in Washington, was arranged by Hermann Biggs, the eminent New York physician and a close follower of Koch, at Koch's suggestion. A special stenographic report of this meeting was made for the Journal of the American Medical Association.[28]

In opening the meeting, Biggs noted that the discussions were informal and private and were not to be communicated to the lay public.[29] Koch opened the meeting with a brief restatement of the views he had already expressed. A heated discussion ensued, in which the main participants were British, French, and American workers. All agreed that the rather technical question of mutability of bacterial types could only be answered by careful laboratory experimentation, most of which had not yet been done. It was also agreed that most humans were infected with tubercle bacilli by the respiratory route, receiving the bacteria from other humans. But it was also firmly the feeling of many present that *some* cases of bovine tuberculosis did occur in humans, and that infection could occur by way of milk. At this point, the following interchange occurred:

Chairman Hermann Biggs: "Is it possible to arrive at any kind of conclusion in a resolution which will put in form the feelings of the members of this conference in regard to this question?"

Robert Koch: "We are dealing exclusively with a purely scientific subject, and the questions at issue can not be solved by any resolution. It is a matter of fact and not of opinion. I for one most energetically refuse to participate in any resolutions that may be passed on this subject."

Dr. Leonard Pearson (U.K): "We have an exceedingly grave responsibility in this matter. It was announced in London in 1901 . . . that Professor Koch took the position that bovine tuberculosis is not transmissible to man, and that measures directed against bovine tuberculosis are of no importance to the public health. . . . Nothing in the world has interfered so much with the control of tuberculosis in cattle as the London address of Professor Koch. . . . I for one should be glad if this meeting . . . would express itself and answer yes or no to each of these questions. If we are not prepared to answer these questions, who is prepared to do so? Where can we find a better jury than right here, and when is there a better time than right now?"

Hermann Biggs: "This [closed] conference was called at the suggestion of Professor Koch to discuss these questions in a scientific and informal way. . . . He has stated that he is not prepared to take up the question of resolutions . . . it seems to me that that is a subject we can not properly consider here unless it is with his full sanction and approval."

Theobald Smith: "If these resolutions are adopted and published we will certainly be asked what specific measures we recommend for the elimination of danger to human beings. . . ."

Professor S. Arloing (from France, one of Koch's strongest opponents on this question): "I think we should all be permitted to express our opinions freely at this conference. . . . It is not my understanding that this is a secret conference. . . . The publication of the results of this conference through proper channels would be a good thing. In voting on these questions I take it that we would vote for the welfare of the public. . . . My dignity would not permit me to be bound to silence on a matter of public welfare by reason of the objection of one invidual."

Professor Koch: "I have no objection whatever to the publication of a verbatim report of this meeting. . . . I propose, then, gentlemen, that after due review by each speaker of the stenographic report of these proceedings that this report be published, but merely as a simple rendition of the words spoken, the facts adduced, and the opinions expressed; but not in the form of any resolution whatsoever . . ."

After brief further discussion, the conference agreed that the verbatim report would be made available as soon as possible. Biggs, perceiving that the debate was about to get out of hand, hastily adjourned the meeting. But the controversy was not over. Despite the fact that the Resolutions Committee of the Congress had established a firm policy that no resolutions of a scientific character were to be passed, the debate spilled over into the Resolutions Committee meeting anyway. Arloing, the chief of Koch's opponents, attempted to secure an unqualified endorsement of the necessity of general pasteurization of milk. Biggs then gave out a statement to the press that said, in effect:

"There was substantial agreement among the delegates present [at the special Willard Hotel meeting] as to both human and bovine tuberculosis being transmissible. The only point at variance was the question of the frequency of human infection from cattle."

The controversy was dealt with by the Resolution Committee in the following terms:

"*Resolved*, That the utmost efforts should be continued in the struggle against tuberculosis to prevent the conveyance of tuberculosis infection from man to man as the most important source of the disease" and

"*Resolved*, That preventive measures be continued against bovine tuberculosis, and that the possibility of the propagation of this to man be recognized."

These resolutions certainly were more anti-Koch than pro-Koch. As one of the foreign delegates said, in a line that was both prophetic and pathetic:

"Dr. Koch isolated the tubercle bacillus; today, science has isolated Dr. Koch."[30]

Here is another view of the Willard Hotel meeting, from a disinterested listener:

The question of human contamination from the bovine tubercular bacilli was hotly discussed, and Prof. Theobald Smith, of Harvard University, wrung from the immortal Koch, who has up to the present time contended that man is not susceptible to the bovine tubercular infection, in a semi-private seance at the Willard Hotel . . . an acknowledgement that his experimentations and demonstrations, too, had proved that man is susceptible to the bovine tubercular bacilli, but that at the present time a public acknowledgement of the fact would be disastrous to himself and his relations to the German government, and that if the matter was to be pressed to the finish he would absent himself from the congress rather than make a public statement of that kind, as it meant so much to Germany, as at the present time their meat products have been prepared under his former idea, and that if he should now renounce his former views and proclaim the views of Professor Smith it would balk the home German meat trade and they would demand American meat, as that is prepared in accordance with the findings and demonstrations of Professor Smith.

Professor Detre, of Budapest, asserted and proved by actual demonstration, before the class of scientists (Professor Koch included) in the Washington Tubercular Hospital, that tuberculosis contracted by a human being from the bovine was distinguishable from that contracted from man, as cases of bovine origin reacted only on bovine tuberculin test. Dr. M.P. Ravanel, of Madison, Wis., and Dr. F. Arloing, of France, were among the warm pursuers of Professor Koch on this question, contending that bovine tuberculosis could be conveyed to the human, and in such cases the lymphatics, bones, joints and peritoneum are likely to be the seat of infection, while pulmonary consumption was more than likely to be from the human tuberculosis.[31]

Koch's influence in America

This debate, and its outcome, reveal clearly the importance with which Koch and his opinions were held in the United States. Although Koch's reputation had undoubtedly been tarnished by the tuberculin fiasco, Koch's sway in America was still considerable. His influence is not surprising. Most of the important medical researchers of this era had either studied with Koch or with other German scientists who had been trained by Koch. It was the Koch canon, after all, that they all were following. It is a bitter realization that Koch's position at the Washington meeting actually became the rallying cry of those who wanted to deny the importance of milk as an agency of tuberculosis transmission. This was a time when the movement for milk pasteurization was very strong, but the "anti-pasteurizers" were still quite vocal. However, after the Washington meeting, Koch's influence in America was virtually eliminated, as the U.S. medical and public health establishment proceeded to so-lidify their agenda for the control of infectious disease. And one of the bulwarks of this agenda was the universal pasteurization of milk. Indeed, a pasteurization temperature high enough to kill the tubercle bacillus was chosen.

In analyzing Koch and the Washington meeting many years later, Theobald Smith made the following statement:

"In training his guns continually against the bacteriological super-stitions of his day, Koch naturally became to a certain degree a victim of the recoil in going too far in insisting on the stability of form and function among bacteria."[32]

After the festive occasion of Robert Koch's first visit to America in the spring, it is bitter to realize how devastating his second, and last, visit to the United States must have been.

Return to Berlin

On 21 October 1908 Robert Koch returned to Berlin from the United States. He immediately initiated new studies on tuberculosis, but, his health failing, they were never completed.

Robert Koch died on 27 May 1910 at Baden-Baden, a victim of a serious heart attack. He was 67 years old. His body was cremated and the remains were deposited in a mausoleum in the west wing of the Robert Koch Institute for Infectious Diseases.[33]

22

An Assessment of Koch and His Work

Although Robert Koch is dead, what he has done will never die.

—PAUL EHRLICH[1]

Robert Koch was one of the most influential and dedicated medical researchers of the nineteenth century. More than anyone, his work placed medical research on a firm scientific footing. His contributions, taken within the broader framework of German science, were so numerous and important that one cannot even imagine how medicine would have developed without Koch. Although he did not even begin to do formal scientific research until he was 37 years old, once he had obtained a full-time post, he carried out his studies with a single-minded fervor that left no room for idleness. Koch's motto, *nunquam otiosus* (never idle) expressed in a nutshell the essence of his life.

Although in his later life Koch was considered by strangers to be cold and authoritarian, among the closed circle of his scientific friends and associates, he was noted for his warmth and humor. His close friend and colleague Paul Ehrlich placed him "among the few princes of medical science". At the start of his career he was a popular and capable family physician, highly sought after by patients. He was noted for his calm, competent manner at the bedside, his thorough dedication to excellence in medical practice, and his willingness to work long hours and to travel long distances for house calls. In his youth he was eager, helpful, enthusiastic. He hated pomp and circumstance. Later in his life, he was showered with honors, which he accepted with modesty. However, he was passed over for the Nobel Prize four times in favor of (mainly) lesser candidates, a reflection of his widespread unpopularity among the medical community.

After a late start, Koch rose rapidly to the forefront of German and international science. Only 9 years after his first paper was published (in 1876), he was appointed (in 1885) Professor at the University of Berlin. And then, 6 years later, he resigned his professorship amid widespread criticism of his tuberculin work. Although he remained in good favor with the Prussian government, who built two major research institutes for him, Koch was increasingly isolated for the last 20 years of his life, travelling away from Berlin on extended research expeditions.

In his later years, Koch appeared to foreigners to be throughly German in character (Figure 22.1), a senior civil servant accustomed to having assistants at his bidding. Yet he actually relished doing his own research work, and even in old age planned and carried out difficult and complex research programs. He was accused of pugnacity, arrogance, failure to give credit for ideas borrowed from others, and reluctance to admit mistakes. All of these criticisms were certainly true. Is this behavior understandable in one who was a true pathfinder?

Although Koch was frequently suspicious and aloof among strangers, his friends and colleagues saw him as a warm and friendly man. Paul Ehrlich described how it was Koch who, at a difficult time in Ehrlich's career, took him in and gave him a position: *"Sie können hier machen, was Sie wollen"* ("You can work here on whatever you want").

> Among my most treasured memories . . . was the *Referierabende* [evening journal club] that took place under his direction at the Institute. He himself reviewed the most important papers. And in almost any subject, he was able to clarify and explain. He was not fond of congresses and large meetings. But for any who had the chance to experience it, it was wonderful to be able to listen to him talk in the narrow circle of his friends and coworkers. In such company, he spoke simply, candidly, and his words imparted a breath of fresh air to whatever subject was under discussion.[2]

Koch the scientist

Robert Koch was an experimenter *par excellence*. Careful and patient observation, hard work, and keen insight were the tools of his trade.

Figure 22.1   *Robert Koch in his 60's.*

He developed new techniques as he needed them and adapted old techniques to new uses. No one has made greater contributions to the development of bacteriological technique than Koch. When working on a difficult problem, he did not give up easily. *"Nicht locker lassen"* ("Never let up") was a favorite exhortation to himself and his associates.[3] Yet, if he became convinced that a line of research was doomed to failure, he readily abandoned it. He had the mark of the great scientist: a concentration on *do-able* research.

Koch's early background in natural history played an important role in the initial development of his research. He was an expert with the microscope, and more than anyone else was responsible for the introduction of the homogeneous immersion lens and the Abbe condenser into biological research. But microscopy alone was not enough. One needed to know *how* to prepare materials for the microscope. And in this, Koch was a master. The staining and slide techniques that he developed for bacteria are still the standard techniques, over 100 years later.

In the early years, many people had trouble repeating Koch's work. He was vilified and castigated; his veracity was doubted. In certain circles he was attacked with a passion. This is the lot of the true pioneer. His unruffled exterior belied the fact that Koch took these criticisms hard. It was not Koch's fault, of course, that his work could not be repeated, but the fault of his critics. Koch's certain knowledge that his critics were wrong led him to become combative and hostile. He was exceedingly opinionated and unyielding when his own ideas were attacked. Although in his youth he had been generally correct, in later years he was sometimes wrong, although unable or unwilling to admit it. During the tuberculin work, he was so beleagured that he left Germany to avoid criticism. And his erroneous ideas about bovine tuberculosis were not only pigheaded, they were dangerous, being responsible for a serious delay in the introduction of appropriate control methods for tuberculosis in cattle, a bad breach of the public health.

Koch and bacteriology

Although in his later years he made useful and important contributions to tropical medicine and parasitology, Koch's main contributions were in the field of bacteriology, especially as it applied to medicine. It can almost be said that Koch *created* the field of bacteriology. Although today bacteriology is a broadly based member of the biological sciences, in Koch's day bacteriology arose as the science of infectious disease studies. Certainly, medical research dominated the field of bacteriology until the end of World War I (when agricultural and general bacteriology finally became established). Koch's influence on medical microbiology can be most effectively demonstrated by reference to the table which summarizes the fruits of Koch's concepts and methods. Between 1876, when Koch published his first work, and the turn of the century, most of the major bacterial pathogens were isolated and characterized. Note that most of the discoverers were German or under German influence.

Koch's Legacy: The Discoverers of the Main Bacterial Pathogens

Year	Disease	Organism	Discoverer
1877	Anthrax	*Bacillus anthracis*	Koch, R.
1878	Suppuration	*Staphylococcus*	Koch, R.
1879	Gonorrhea	*Neisseria gonorrhoeae*	Neisser, A.L.S.
1880	Typhoid fever	*Salmonella typhi*	Eberth, C.J.
1881	Suppuration	*Streptococcus*	Ogston, A.
1882	Tuberculosis	*Mycobacterium tuberculosis*	Koch, R.
1883	Cholera	*Vibrio cholerae*	Koch, R.
1883	Diphtheria	*Corynebacterium diphtheriae*	Klebs, T.A.E., Loeffler, F.
1884	Tetanus	*Clostridium tetani*	Nicholaier, A.
1885	Diarrhea	*Escherichia coli*	Escherich, T.
1886	Pneumonia	*Streptococcus pneumoniae*	Fraenkel, A.
1887	Meningitis	*Neisseria meningitidis*	Weischselbaum, A.
1888	Food poisoning	*Salmonella enteritidis*	Gaertner, A.A.H.
1892	Gas gangrene	*Clostridium perfringens*	Welch, W.H.
1894	Plague	*Yersinia pestis*	Kitasato, S., Yersin, A.J.E. (independently)
1896	Botulism	*Clostridium botulinum*	van Ermengem, E.M.P.
1898	Dysentery	*Shigella dysenteriae*	Shiga, K.
1900	Paratyphoid	*Salmonella paratyphi*	Schottmüller, H.
1903	Syphilis	*Treponema pallidum*	Schaudinn, F.R., and Hoffmann, E.
1906	Whooping cough	*Bordtella pertussis*	Bordet, J., and Gengou, O.

Despite the strong medical emphasis of Koch's bacteriological work, it was in the beginning firmly rooted in natural history. Koch's early schooling by his uncle Eduard provided a framework and an understanding of living organisms that made it possible for Koch to approach bacteria as unique living systems. To an important extent, Koch's work can be considered as applied ecology, since it was dominated by an interest in how bacteria spread from host to host. Koch's most important contributions to medicine were in the fields of public health and hygiene of acute infections, fields which he almost single-handedly started. Although Koch never attempted to understand the physiology of bacteria, his strong naturalist bent (which remained with him throughout his life) kept him thinking along lines that were exactly appropriate for public health research. Others, Pasteur and Ehrlich especially, would make the key contributions to the control of infectious disease in individuals, but Koch's work was central for the control in populations.

Koch approached his first work, on the life cycle of *Bacillus anthracis*, as a problem in natural history, but he quickly discovered the practical significance of his knowledge for the control of the disease. He never carried out any research that did not have, at its root, a practical question, generally medical.

Koch showed the medical world how to use a microscope properly. Even though he had no training in microscopy or physics, as a careful observer he developed the most effective means for observing the small creatures that were his main interest. He spent a year trying to obtain useful photomicrographs of bacteria, being convinced that only with pictures could the objective reality of bacteria be communicated to other workers. The photomicrographs which he took with primitive and homemade equipment were superb. They are almost unsurpassed to this day.

Although Koch's self-confident arrogance has been alluded to frequently, this was apparently a personality trait which developed later in life. There was no arrogance in the eager young physician who went to visit Ferdinand Cohn in order to show him his work on the anthrax bacillus. When he completed his anthrax work, he was wracked with self-doubt. Only after he had presented his work to experts was he certain that he was on the right track. However, once his work was acclaimed (perhaps too highly), Koch gained supreme self-confidence. Never again did he outwardly show any sign of self doubt or uncertainty about his work.

Koch developed the important plate technique primarily to permit the isolation in pure culture of individual species of bacteria. Again, this was a natural history approach. The necessity of pure cultures had been discussed by Jacob Henle and the project had been attempted enthusiastically by Edwin Klebs, but it was Koch who found the way to purify cultures accurately and reproducibly. The ramifications of this simple technique were so widespread that the whole field was turned on its end. The plate technique not only permitted pure culture isolation, it permitted *quantification* of bacterial numbers. Sterilization and disinfection techniques could be put on a quantitative basis. Assessment of the microbial content of air, water, and clinical samples became possible. Other outcomes were to appear later, of which the most important was the use of the plate technique for studying the *genetics* of bacteria. Biotechnology as we know it today could not have been conceived without Koch's plate technique.

Showing his willingness to tackle *big* problems, Koch trained his new techniques on the biggest medical problem of the day, tuberculosis.

And in an amazingly short time, he had solved it. It is still inconceivable that it took Koch only nine months to detect, isolate, and characterize the tubercle bacillus. His success was due to a combination of accident, perseverance, good luck, and keen observation. But no one who had not trained himself so thoroughly in microscopy and bacteriology could have done it. His staining method and plate technique provided the necessary tools.

It is hard to appreciate, at this late date, what a sensation Koch's isolation of the tubercle bacillus caused. At that time, tuberculosis was the number one infectious disease of the human race. And the pathology of infection with the tubercle bacillus was so complex and difficult that it was not even certain that the different manifestations all had the same underlying cause. Koch's discovery, in particular the special staining procedure which Ehrlich rapidly improved, made it possible not only to trace the organism through the body, but to *diagnose* the infection. The slow-growing tubercle bacillus presented numerous experimental difficulties, which Koch overcame in masterful fashion. Koch's first lecture on the tubercle bacillus, given at the Physiological Society in Berlin, stunned his audience. Soon, the medical world of Europe was at Koch's feet.

But Koch did not rest on his laurels. Another *big* disease was threatening—cholera—and Koch quickly turned to its study. Carrying out bacteriological work under extremely difficult conditions in Egypt and India, within a year he had isolated the causal agent, the cholera vibrio (comma bacillus). These two discoveries, the tubercle bacillus and the comma bacillus, coming so quickly one after the other, established Koch's reputation for all time. Honors were heaped upon him. His utterances were carefully recorded and repeated. A *Koch myth* developed: the great Koch could do no wrong. If the great Koch said it, it was true. Unfortunately, even Koch had difficulty distinguishing the myth from the reality. Although he made many important contributions after his cholera work (among which, the discovery of tuberculin and typhoid carriers were the most noteworthy), never again would his work progress as ably as in those few years at the Imperial Health Office right after he came to Berlin.

Soon Koch lost *control* of his own work, swept up as he was in the administrative duties of his position. He no longer had time to work effectively with his own hands. But although he himself became less and less involved in effective research, the field of bacteriology was established, and Koch's many co-workers and followers readily elaborated upon his initial contributions to the field. By the end of the nineteenth century, bacteriology was a major, growing discipline (see the table).

Koch and medicine

Koch was trained as a physician and he practiced medicine for over ten years before he began to do serious research. Throughout his life, his research focused almost exclusively on medical problems. Koch's contributions to medicine were so numerous and fundamental that we can recognize two major eras, *before Koch* and *after Koch*. All of our advances in the control of infectious disease build on Koch's seminal work on the germ theory of disease. But Koch's major contributions to human health were in the field of public health and hygiene. His forays into therapy and immunization, notably in the tuberculin work, were disastrous and seriously flawed.

There was a marked contrast between the French and the German approaches to infectious disease. Pasteur and the French school developed treatments for *individuals*, whereas Koch and the German school developed approaches for the control of infectious disease in *populations*. The strong central government that dominated Germany at that time was probably a major factor in the success of this approach. Laws could be established and regulations promulgated that would have marked impact on the spread of disease. Advances in food and water bacteriology were built on the foundations which Koch laid. Hygiene, public health, social solutions to infectious disease—these are the things for which Koch can be remembered.

The militaristic approach to research of the Koch group was frequently noted by Koch's contemporaries outside Germany. In those ancient days before World War I, the military was all powerful in Germany, and many of Koch's assistants were or had been officers in the German army. They were employees of the Empire. When Koch's coworkers went into the field, they went wearing uniforms. Koch was extremely loyal to his Emperor and to his government.

Koch played a major role in writing the government regulations for public health. He was a member of every German commission, attended all the international congresses for sanitation and hygiene, wrote numerous drafts of regulations and guidelines. In the three volumes of his collected works (*Gesammelte Werke*), one volume contains a vast number of such documents which Koch wrote or participated in writing.

Public health as we know it today began with Koch, and the German approach is especially engrained in the sanitary regulations of the Anglo-Saxon and Scandinavian countries.

Koch's other major medical contribution was in pathology. Although never a tissue pathologist like Virchow and Cohnheim, Koch's important contributions to the bacteriology of infectious disease established a new era in pathological research. Indeed, microscopic pathology was at the root of Koch's research on infectious disease. The first of Koch's postulates: *The suspected pathogen must be constantly present when the disease is present* can only be fulfilled by means of microscopic pathology.

Although Koch's discovery of tuberculin was a major milestone in the field of immunology, Koch himself never understood the significance of his discovery. It was left to others, much later, to explain the Koch phenomenon (as it was called) and assess its significance for infection and immunity. In Koch's time, tuberculin reactivity came to serve primarily in the *diagnosis* of the disease. In this regard, it was an important discovery, but most of the advances here were made by others. Koch's discovery of tuberculin was almost his downfall. The tuberculin fiasco would have destroyed a lesser man. Actually, blame for the disaster cannot be placed completely with Koch. Tuberculosis is such a complex disease, and tuberculin reactivity so arcane a phenomenon, that a satisfactory solution during Koch's lifetime would hardly have been possible. The times were not right.

But Koch's work on tuberculin shows another side of his character: the willingness to tackle *big* problems. Bacteriology, after fifteen years of intense work, had not led to any *practical* medical accomplishments. Koch saw the promise of tuberculin as justification of bacteriological research. He was wrong. But the main criticism of Koch here was not the work, but its premature announcement and the veil of secrecy he tried to maintain. A decade earlier, he would have been able to complete his tuberculin work before announcing it to the public, but now things got out of control. There is no way Koch could have anticipated the absolutely insane hue and cry that arose over the tuberculin announcement. However, Koch's unwillingness to admit his error and his use of the tuberculin discovery to improve his own position were reprehensible. Although there is no evidence that Koch profited financially from the sale of tuberculin, he did use its discovery as a lever to obtain a new institute and to obtain a marked increase in his own salary. Koch apparently never lived luxuriously, but he clearly loved the finer things that money could buy. Although Koch's reputation remained firm within Germany, after tuberculin he was never looked upon in the rest of Europe with the same reverence.

Interestingly, if Koch had worked on a less important disease, such as tetanus or diphtheria, he might have developed a successful treatment. Bacteriology's first major triumph in the field of infectious disease came not from Koch, but from Behring, one of Koch's assistants and a much lesser figure. It must have been bitter to Koch that when the Nobel Prize in medicine was established, the *first* prize went to Behring rather than himself. Without Koch before him, Behring's work would have never been possible.

Koch showed little or no interest in chemistry. He apparently had little interest in the fundamental processes of bacteria, and despite the fact that his pure culture technique was essential for studies in bacterial physiology, he made no contributions to this field. He never studied nonpathogenic bacteria, except as they might become contaminants in his pure cultures. He studied bacteria because, and only because, they caused disease. Certainly Koch's attitude had strong influence on how academic bacteriology developed, both in Germany and elsewhere. Bacteriology has been, and still often is, looked upon primarily as a medical subject, only peripherally connected to biology. Even in modern times, nonmedical developments in bacteriology and especially in molecular biology have come primarily from *outside* academic bacteriology and microbiology departments.[4]

Finally, one cannot complete a discussion of Koch's influence in the field of medicine without considering how his discovery of the tubercle bacillus affected the incidence of tuberculosis in humans. Figure 22.2 shows how the incidence of tuberculosis has changed since Koch's discovery of the tubercle bacillus. The disease rate has dropped steadily, albeit not dramatically, since early in the nineteenth century, but it is difficult to perceive that *any* medical discoveries have influenced this drop in rate. Certainly the discovery of the tubercle bacillus had little effect on the progress of that curve, nor did the discovery of tuberculin or any vaccination procedures. Even streptomycin, one of the most effective of therapeutic measures for tuberculosis, has affected primarily the incidence of serious cases. Tuberculosis has quietly gone away, but not because of *specifics*, either chemical or immunological. One must conclude that tuberculosis was conquered primarily by public health measures and by a general improvement in the health of human society.[5]

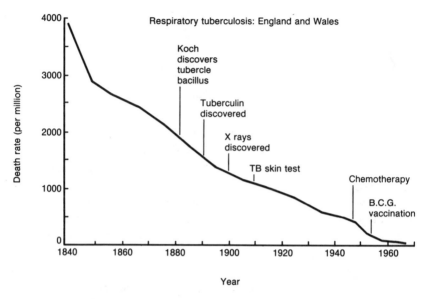

Figure 22.2   *Two centuries of decline in tuberculosis incidence. The mean annual death rate in England and Wales is shown along with key scientific events affecting our understanding of the disease.*

Koch the man

Robert Koch's letters tell of a friendly, enthusiastic man who loved his daughter, had broad interests, and worked hard for the advancement of his colleagues and friends. His approach to women was standard nineteenth century male chauvinism. Women to him were vassals, servants, sources of amusement and entertainment, but hardly equals. He was very careful to ensure that his daughter was educated in home-making, the practical arts, and languages, but not in anything "masculine". He took an interest in his grandchildren, but in a remote way.

He was an ardent mountain climber and alpinist, in the heroic days of these activities before vibram soles and down jackets. He played chess enthusiastically, following the chess world through periodicals and books. He loved nature and revelled in the wonders of Africa. One of his real passions was archaeology, in which he even did semiprofessional work in his youth. Throughout his travels he visited the famous archaeological sites. He loved seeing new things, but showed no interest in politics. Religion never entered his life. Although he was extremely fond of his daughter, he showed no passion for his first wife, and, only briefly, for his second wife.

The relationship between Koch and Ferdinand Cohn is an interesting one. Cohn had a botanical approach to bacteriology but recognized the significance of bacteria as agents of disease. Cohn's promotion of Koch and his work was central to Koch's success and to his rise in the scientific hierarchy. Yet, despite Cohn's friendliness and enthusiasm for Koch's work, once Koch went to Berlin, there is no evidence that he remained in contact with Cohn. Cohn himself did little bacteriological work after 1880, although he continued to publish the work of his students and colleagues in his journal (*Beiträge zur Biologie der Pflanzen*). After Koch became famous, a myth developed that Koch had been a student of Cohn's and that his ideas owed much to Cohn's influence. Cohn himself, in 1890, wrote and dispelled this myth, emphasizing that Koch had already completed his work when he came to Breslau to demonstrate it.[6] However, without Cohn, Koch's way would have been exceedingly hard. Was Koch grateful? It is hard to know. They seem never to have corresponded after Koch left Wollstein.

In later years, Koch had many detractors, as do most great men. But he also had many champions. Several quotations can confirm that Koch was a multifaceted man, who was perceived differently by different people.

Those were the days in which the tubercular bacillus discovered by Robert Koch was making its way in the world and tuberculin treatment was becoming fashionable. Koch's tuberculin was bought up by the Hoechster Farbwerke for a million marks, a very large sum in those days. Its possession enabled Koch to divorce his first wife and enter into a new matrimonial venture with a plump, blonde and most attractive young lady from the stage. I have often wondered whether the undoubted counter-attractions of life with this young woman (much younger than himself) had anything to do with the fact that Koch released his valuable discovery for general use before carefully seeing it through the requisite long period of tests. In any case, that is what he irresponsibly did, with the result that his specific very quickly got into the hands of incompetents who used it without discrimination, causing a great deal of avoidable damage.[7]

Many felt that the weight of Koch's authority in his domain was at times too great, especially in its influence on Government, and, unless Koch expressed a favourable opinion of a discovery, it was liable to be received with doubt by German men of science. Moreover, when Koch or members of his school had been given the opportunity of confirming the work of others, there was a decided tendency to arrogate to themselves undue credit in discovery. This resulted in ill-feeling that was none the less bitter because it did not always find its way into print. Men feared to stand up against Koch because they felt his influence, if directed against them, might injure them in their career or lead to the loss of a coveted post. Under the circumstances, Koch was in a position to push others aside and on occasion he did so. Thus, to my knowledge, when an obscure but observing young veterinarian in German East Africa had found *Piroplasma bigeminum* in cattle suffering from redwater and had shown the parasites to Koch, we only heard of Koch's work thereon afterwards. When Metchnikoff brought malaria blood-films from Russia in 1887, and demonstrated these to Koch, he was treated, as he told me himself, with scant courtesy, for Koch apparently still disbelieved in the existence of the malaria parasites that Laveran had discovered in 1880; at any rate Koch denied that Metchnikoff's slides showed malaria parasites. The latter, had, however, demonstrated his excellent preparations to Professor Flügge and to me some days previously in Breslau and we had seen the parasites. This experience certainly hastened Metchnikoff's journey on to Paris. . . . Although it is unpleasant to dwell on this aspect of Koch's character I have felt it incumbent upon myself to do so with a view to giving a true picture of the man, not as he might have been but as he was. The instances I have cited might be multiplied, but they will suffice to show that, great as he was, Koch had his weaknesses. I presume that constant appeals to him for his opinion may have adversely affected his judgement on some matters, in which case there are obvious disadvantages in being too great an authority in the eyes of one's fellow-men.[8]

This quotation recalls to mind Koch's own experience when he tried to show his slides to Virchow in Berlin (see Chapter 8). Does history repeat itself? On the other hand, the following experience of a visitor to Koch in his later years is quite the opposite:

An introduction to Prof. Rabinowitsch, a Russian bacteriologist working in the Koch Institute, made it possible for me to meet Prof. Koch himself. . . . [Prof. Rabinowitsch] greeted me pleasantly and without ado said she would *send for* Prof. Koch, who was downstairs. I protested at such an imposition upon the Geheimrath Koch as to request him to climb a staircase to see an American stripling, but with characteristic directness of action she sent a "diener" or assistant to get him. I was mortified, and when he shortly appeared, stumbled into an apology with what little German I could command for the occasion. He was a very quiet, white haired man, without indication in his face of either pleasure or disgust at my visit. Behind large spectacles, he was the typical plodding German, with little formality—but only formal to strangers who caused him to climb two flights of stairs at the behest of a slight, black-eyed, and determined Russian woman!

After my explosion of pigeon-English-German (which must have secretly amused him) he responded: "You need not speak in German; I speak English." This was a relief, even though his English was Germanized. Then he said, quite without impatience:

"What would you wish here to do?"

Only to ask him a few questions about tuberculin, and to have one of his assistants show me something about the tuberculosis experiments.

"I will take you to Dr. Neufeld, my assistant," he said calmly, and in halting words. On the way as we walked downstairs he stated that he considered tuberculin still the best treatment for tuberculosis, and that the latest tuberculin contained no living germs, as it was sterilized by glycerin. . . .

I mentioned a certain New York doctor purporting to be "Dr. Koch" had exploited himself and had an "institute" in which he gave a "cure" for tuberculosis.

"He is ein Schwindler! How can I him stop?" he said with some heat. . . .

After a few other questions on minor points, partly answered in German, I was politely introduced to Dr. Neufeld . . . Prof. Koch thereupon left me as impressively as he had greeted me, my profuse thanks being accepted with a simple nod. All the assistants in a Berlin hospital or laboratory in those days greeted and parted from a visitor with military bow, known to American students as the "jackknife". It is the peculiar heel-clicking-stiff-back-bow from the hips. . . .

I can truly say that my reception was very courteous and my interview with Koch entirely satisfactory. As it was generally known that Koch was not a man given to cultivating the social graces, I felt that he had been actually cordial to me. One must be flattered to have had Robert Koch speak English to him! Like other discoverers, he was the object of much criticism and jealousy. It is but natural that he should have become suspicious.

Painstaking and thorough by instinct, Koch was a worker rather than a leader of men. This would be the impression a casual visitor would readily receive. I found him without a trace of self-importance, and retain the memory only of a matter-of-fact man without pose or sign of condescension.[9]

Of course, in this case the visitor was of no *scientific* consequence to Koch.

And finally, Koch's personality in his later years is encapsulated in Metchnikoff's own words:

All the biographies of Koch that I know of have clamped down the lid on his family life. This may lead one to believe that it was not worthy of the memory of the great scientist. Therefore, I regard it as my duty to impart to the reader the facts known to me. . . .

The fight against tuberculosis cost Koch a great intellectual effort. Therefore, to be distracted from his labors he was an indefatigable frequenter of Lessing Theatre, not far from his laboratory. There he became enamored of a young, talented and intelligent actress who played minor roles.[10] A romance followed. Koch divorced his first wife and married this young woman. This event unleashed a moral storm, as was to be expected. During the Congress of German Physicians in 1892, where I was present, Koch's marriage was the topic of all conversation. Koch, whose scientific greatness had not as yet been forgiven him, was exposed to the most serious accusations; his romance certainly interested the professors more than all the reports submitted to the Congress. . . .

From the time of its appearance, the first work of Koch inspired in me a feeling of respect. That feeling was changed into out-and-out veneration after reading his first report on the tubercular bacillus. . . . Koch, who was followed by a whole school of young bacteriologists, was at first an enemy of my theory of immunity against infectious diseases. The subjects of research which he assigned to his disciples were directed against me. In 1887 at the Congress of Hygiene in Vienna, I met his chief assistant who informed me that Koch was anxious to see the preparations connected with my last work on recurrent fever; that he had asked to have them sent to him. Naturally, I welcomed this idea and so I replied that instead of sending them I would bring them myself. Several well-known Munich bacteriologists advised me against it, for they were convinced that Koch would show me out; that he would fail to see in these preparations what I found in them and that he would be able to assert henceforth that he had established the lack of foundation in my conclusions with a full knowledge of the facts. However, I did not heed this warning and betook myself to Berlin. Arriving at the Institute of Hygiene where Koch was teaching, I met his assistants and pupils. After announcing my visit to Koch, the meeting time was set for the day after. In the meantime I had taken out my preparations and shown them to the young assistants. They all stated unanimously that what they observed under the microscope undoubtedly confirmed my conclusions. Much encouraged by that testimony I went to Koch's laboratory the following day accompanied by his chief assistant. There, seated at the microscope, I perceived a rather elderly man. [Koch was only 44 years old in 1887!] He was rather bald and his thick, bushy beard showed as yet no traces of gray. His handsome face had a serious and almost haughty expression. His assistant with a show of deference made the announcement that I had come to keep the appointment fixed for that day and that I was anxious to show him my preparations. "What preparations?", he exclaimed in a cross tone of voice. "I've told you to prepare everything necessary for today's lecture and I see that many things are still missing!" The assistant excused himself humbly and again pointed me out to Koch. Without shaking hands with me, the latter told me that he was very busy at that moment and that he could devote only very little time to my preparations. Some microscopes were hastily brought together and I managed to show him what to me seemed the most convincing details. "Why, then, have you made use of a violet coloration when a blue one would have been much better?" I explained the reason to him but this failed to soften him. After a while he got up and declared that my preparations were far from convincing and that they failed to substantiate my point of view. Much galled by these words and by Koch's behavior, I replied that a few moments were evidently not sufficient for him to appreciate the fine points of the preparations, and I asked him for another lengthier meeting. At this, the assistants and students who the day before had been on my side, made common front with him.

During the second meeting he showed himself somewhat more conciliatory. After attempting to refute my arguments, he determined to recognize the evidence, but finally concluded with the following words: "You know, I am not a specialist in microscopical anatomy; I am a hygienist. Consequently, it is entirely indifferent to me whether the spirilla are within or without the cells." Upon this I took my leave.

It was not until nineteen years after that meeting that Koch publicly confirmed in scientific journals what I had a long time before stated regarding my preparations. Much had happened in the interval: Koch had already published his researches on tuberculin; he had been persecuted and had suffered many mortifications. From a director of the Institute of Hygiene he had become the director of the Institute for Infectious Diseases with his own private clinic. During one of my trips to Berlin, in 1894, I went to see Prof. Pfeiffer, one of the heads of his laboratory . . . "Do you know that Herr Privy Councillor is here?" he said. I remembered too well the cold reception I got from Koch during my first visit to express the desire of again being introduced to His Eminence. Almost immediately Pfeiffer left and soon returned with Koch. He was entirely changed. He very kindly showed me his clinic and his patients. He entered upon minute details of tuberculin treatment and sharply criticized the doctors who failed to avail themselves of it. Finally he invited my wife and myself to dinner and presented us to his wife. The memory of our first meeting was completely forgotten.

During the period that followed Koch made long voyages to the Indies, to Africa and elsewhere. I saw him again ten years later, first in Berlin, then in Paris, which city Koch was anxious to show to his wife. Mme. Koch insisted on visiting the theatres and the famous artists. I feared that these theatre parties and evening entertainments would tire him out as he was already in his sixties. When on the night before their departure Mme. Koch evinced a desire to see the Montmartre cabarets, I suggested as a guide a young doctor who would gladly undertake such a task. Koch, however, insisted on accompanying his wife and the two went to see the absurd shows of Montmartre. Koch enjoyed the Parisian restaurants and mocked at my persistence in following the rules of hygiene, calling it pedantic. . . .

Although Koch's stay in Paris was not for scientific purposes, yet he was shown everything that might interest him in his field. The reception accorded him at the Pasteur Institute surpassed that of all the crowned heads. The staff met him at the library and wished him a hearty welcome with a salvo of unanimous applause. Koch visited the laboratories, the stables and everything else, but showed greatest interest for the technical details. He took note of the slightest improvements in procedure—the blood-letting of horses, injections, etc. I introduced him to Curie who showed us his experiments with radium and its emanations.

During his very satisfactory stay at Paris, Koch had the occasion to visit several museums. While visiting the Louvre he evinced a profound knowledge and a fine taste for paintings. As a whole, he was far from being the narrow-minded specialist one might imagine on reading certain of his treatises; on the contrary he was well versed in many fields of knowledge. In philosophy he was a follower of Mach. Somewhat later he sent me as a souvenir one of the works of that philosopher. We parted good friends.

Koch and I met for the last time in the summer of 1909. I found him in his laboratory, immersed in his researches on tuberculosis which he attempted to cure by new preparations of tuberculin. He looked very well, full of life. And nothing indicated his approaching end. He was dead eleven months later.[11]

Final Words

Robert Koch's story can be an inspiration to all who are fascinated by the interplay of science and medicine. His story exhibits the importance of diligence and persistence in scientific research. Koch's life tells us that even without "credentials" one can, through supreme effort, complete important scientific research and, eventually, achieve the recognition of one's peers.

Could a Koch happen today? Are science and medicine so complex, expensive, and technical that lone individuals, working in isolation, would be unable to make significant accomplishments? Is it the case that today only those who grow up inside the scientific establishment could ever hope to achieve anything significant? I believe that a Koch *could* happen today. Not, of course, within one of those firmly established fields of science or medicine where significant work can only be done with lots of money and vast amounts of equipment and space. But in those offbeat fields that are orphans of the establishment, significant and important research awaits to be done by those who, even untrained, would do research. In Koch's day, bacteriology was just such an off-beat field. Its eternal significance was never suspected by those in the medical or biological establishment.

It has been said that science continually opens up new fields to view. Many new fields, whatever they may be, await the coming generations of Robert Kochs!

Chronology

1843

Robert Koch born, December 11

1848

Begins elementary school at age 5

1851

Begins studies at gymnasium

1862

Graduates from the gymnasium at age 19
Enters the University of Göttingen

1866

Completes university examinations, January 13
Attends Virchow's lectures in Berlin, January–March
Passes medical examinations, 12 March
Engaged to Emmy Fraatz
Moves in June to Hamburg for first medical duties
Moves in September to Langenhagen as physician for mental hospital

1867

Marries Emmy Fraatz, 16 July

1868

Loses position at Langenhagen
Moves in June to Niemegk as practicing physician
Daughter Gertrud Koch born 6 September

1869

Moves in July to Rakwitz near Posen

1870

Serves in German army in Franco-German War

1871

Released from Army to attend his dying mother

1872

Takes examination for position as District Physician
Becomes District Physician in Wollstein

1873

Begins first research on anthrax
Abbé condenser described

1875

September–October, takes long scientific trip; first insight into the medical research community
Inoculates rabbit 23 December, experiment leading to discovery of endospores

1876

April–May, visits Ferdinand Cohn to demonstrate anthrax life cycle
Famous paper on anthrax published, December

1877

Koch perfects techniques for microscopy of bacteria
First photomicrographs of bacteria
Koch demonstrates anthrax to Burdon Sanderson, October 15
Nägeli's book on pleomorphism published

1878

Sees first oil-immersion lens in Jena
Visits Leipzig and Berlin to demonstrate his techniques
September, presents paper at meeting in Kassel

1879

Abbé and Carl Zeiss supply first oil-immersion lenses, developed especially for bacteriology

July, Koch moves to Breslau to become District Physician

October, Koch, unhappy, returns to Wollstein

1880

April, Koch approached about position in Berlin

July, Koch moves to Berlin as staff member of the Imperial Health Office

1881

The famous "Koch plate technique" published

August, Koch demonstrates his techniques to Lister and Pasteur in London

18 August, begins first work on tuberculosis

1882

January, Koch promoted to Senior Medical Officer

24 March, the famous address on the tubercle bacillus to wide acclaim

10 April, the paper on tubercle bacillus published

May, Ehrlich's paper on acid-fast staining published

Fannie and Walther Hesse develop agar

Koch attacks Pasteur's work

Thuillier conducts successful tests of Pasteur's anthrax vaccine in Germany

September, Pasteur and Koch confrontation at Geneva

25 December, Pasteur publishes open letter to Koch

1883

The "postulates" first formulated

The German Exposition of Hygiene and Public Health opens in Berlin

Loeffler describes the causal agent of diphtheria

16 August, Koch leaves Berlin for Egypt to study cholera

18 September, Thullier, from the French cholera team, dies of cholera in Alexandria

November, Koch travels from Egypt to India

11 December, arrival in Calcutta

1884

February, the cholera vibrio isolated in pure culture

May, return to Berlin with high acclaim

June, Koch receives high honors from the German government

July, a major cholera conference held in Berlin; all of Koch's claims confirmed

1885

Koch appointed Professor of Hygiene at the University of Berlin, and director of the Hygiene Institute; resigns from the Imperial Health Office

May, Koch and Flügge establish a new journal, Zeitschrift für Hygiene

July, Hygiene Institute opens

November, Koch gives first lecture as a professor

1887

Gertrud Koch engaged to Eduard Pfuhl, one of Koch's assistants

Koch receives honors and awards from many countries

Koch's personal research stagnates while he writes lectures and travels on extensive vacations

1890

Koch returns to research on tuberculosis

4 August, Koch's announcement that he has found a cure for tuberculosis; the first report of tuberculin

August–December, massive excitement and controversy over Koch's new "secret" remedy for tuberculosis

October, Koch resigns his professorship to spend full-time on tuberculin

A new institute is announced for Koch, the Institute for Infectious Diseases

Koch receives widespread awards and acclaims throughout Germany

4 December, Behring and Kitasato publish their epoch-making paper on diphtheria and tetanus antitoxin

1891

His personal life and scientific research a shambles, Koch leaves for extended Egyptian vacation, refusing to return until the Prussian government has appropriated an operating budget for his new institute

May, the Prussian parliament approves the budget for Koch's new institute

July, the new Institute for Infectious Diseases opens

Koch publishes papers on tuberculin, finally describing what it is and how it is made

1892

Cholera outbreak in Hamburg permits Koch to demonstrate the importance of water filtration in control of cholera and typhoid

1893

Robert and Emmy Koch are divorced

13 September, Koch, 50, marries the twenty-year-old Hedwig Freiberg

Behring develops effective diphtheria antitoxin

1894

Behring leaves Koch's laboratory and sets up own institute

1896–1900

Koch makes numerous research visits to Africa, to study diseases of cattle for the British and German governments; to India to study plague; to Italy to study malaria

1900

The Robert Koch Institute for Infectious Diseases opens in the north of Berlin

1901

Behring awarded the Nobel Prize for medicine for his discovery of antitoxin therapy, the first Nobel Prize awarded in medicine

1902

Research on typhoid fever in Trier leads to Koch's discovery of the carrier state in humans, an important public health concept

Trip to Rhodesia for the British government to study rinderpest

1904

October, Koch officially retires from government service

Koch travels extensively in Africa, studying tropical diseases

1905

12 December, Koch finally awarded the Nobel Prize, having been passed over earlier in favor of von Behring (for antitoxin), Ronald Ross (for discovery of the malaria parasite), Niels Finsen (for use of light therapy in disease treatment), and Ivan Pavlov (for work on conditional reflexes)

1906–1907

Koch back in Africa to study sleeping sickness

1908

April, Koch in New York, receives honorary banquet

June, Travels to Japan via Iowa and San Francisco

September, Koch in Washington for International Tuberculosis Congress; controversy regarding identity of bovine and human tuberculosis

1909

Declining health

1910

27 May, Koch dies in Baden-Baden at age 67

I. Publications of Robert Koch*

1865

Koch, R. Ueber das Vorkommen von Ganglienzellen an den Nerven des Uterus. Gekrönte Preisschrift der Universität Göttingen. *Gesammelte Werke 2/2*, 806–813, Plates XLI–XLIII.

Koch, R. Ueber das Entstehen der Bernsteinsäure im menschlichen Organismus. *Zeitschrift für rationelle Medizin, Third Series*, 24:264–274. *Gesammelte Werke 2/2*, 814–820.

1876

Koch, R. Die Aetiologie der Milzbrand-Krankheit, begründet auf die Entwicklungsgeschichte des Bacillus Anthracis. *Beiträge zur Biologie der Pflanzen* 2:277–310. *Gesammelte Werke 1*, 5–26, Plate I.

1877

Koch, R. Verfahren zur Untersuchung, zum Conservieren und Photographiren der Bakterien. *Beiträge zur Biologie der Pflanzen* 2:399–434. *Gesammelte Werke 1*, 28–50, Plates II–III (24 Photos).

1878

Koch, R. Review of *C. v. Naegeli, Die niederen Pilze in ihren Beziehungen zu den Infectionskrankheiten und der Gesundheitspflege. München 1877,* and Hans Buchner. *Die Naegeli'sche Theorie der Infectionskrankheiten in ihren Beziehungen zur medicinischen Erfahrung, Leipzig 1871.* Published in *Deutsche medizinische Wochenschrift* 4:7–8 and 18–19. *Gesammelte Werke 1*, 51–56.

*To the extent possible, the citations in this bibliography are to the original literature. All of Koch's papers have been collected in three volumes (*Gesammelte Werke* Volume 1 and Volume 2 Parts 1 and 2 (abbreviated 2/1 and 2/2). For convenience, the page numbers in the *Gesammelte Werke* are also given. Koch also had many unpublished papers and lectures. These are not given in the present bibliography but can be found in the *Gesammelte Werke.*
This bibliography was compiled by Hanspeter Mochmann and Werner Köhler and is used here in edited and abbreviated form with their kind permission.

Koch, R. Neue Untersuchungen über die Mikroorganismen bei infectiösen Wundkrankheiten. Based on his report in the Section für pathologische Anatomie und für innere Medicin der 51. deutschen Naturforscherversammlung zu Cassel. *Deutsche medizinische Wochenschrift* 4:531–533. *Gesammelte Werke 1*, 57–60.

Koch, R. *Untersuchungen über die Aetiologie der Wundinfectionskrankheiten.* Leipzig, F. C. W. Vogel, 1878. *Gesammelte Werke 1*, 61–108, 5 Plates (in *Gesammelte Werke* Plates IV–V with 14 figures).

1881

Koch, Robert. Zur Untersuchung von pathogenen Organismen. *Mittheilungen an der Kaiserlichen Gesundheitsamte* 1:1–48, Plates I–XIV (84 figures). *Gesammelte Werke 1*, 112–163, Plates VI–XIX.

Koch, Robert. Zur Aetiologie des Milzbrandes. *Mittheilungen an der Kaiserlichen Gesundheitsamte* 1:49–79. *Gesammelte Werke 1*, 174–206.

Koch, Robert. Ueber Desinfection. *Mittheilungen an der Kaiserlichen Gesundheitsamte* 1:234–282. *Gesammelte Werke 1*, 287–338.

Koch, Robert, und Gustav Wolffhügel. Untersuchungen über die Desinfection mit heisser Luft. *Mittheilungen an der Kaiserlichen Gesundheitsamte* 1:301–321. *Gesammelte Werke 1*, 339–359.

Koch, Robert, G. Gaffky und F. Loeffler. Versuche über die Verwerthbarkeit heisser Wasserdämpfe zu Desinfectionszwecken. *Mittheilungen an der Kaiserlichen Gesundheitsamte* 1:322–340. *Gesammelte Werke 1*, 360–379.

Koch, R. Entgegnung auf den von Dr. Grawitz in der Berliner medicinischen Gesellschaft gehaltenen Vortrag über die Anpassungstheorie der Schimmelpilze. *Berliner klinische Wochenschrift* 18:769–774. *Gesammelte Werke 1*, 164–173.

1882

Koch, Robert. Die Aetiologie der Tuberculose (Based on a lecture to the Physiological Society of Berlin given on 24 March 1882). *Berliner klinische Wochenschrift* 19:221–230. *Gesammelte Werke 1*, 428–445.

Koch, Robert. Ueber die Aetiologie der Tuberculose. *Verhandlungen des Congresses für Innere Medicin, Erster Congress*, J. F. Bergmann, Wiesbaden, 1882, 56–66. *Gesammelte Werke 1*, 446–453.

Koch, Robert. *Ueber die Milzbrandimpfung. Eine Entgegnung auf den von Pasteur in Genf gehaltenen Vortrag.* Thieme, Leipzig 1882. *Gesammelte Werke 1*, 207–231.

1883

Koch, Robert. Kritische Besprechung der gegen die Bedeutung der Tuberkelbacillen gerichteten Publicationen. *Deutsche medizinische Wochenschrift* 9:137–141. *Gesammelte Werke 1*, 454–466.

1884

Koch, Robert, G. Gaffky, and F. Loeffler. Experimentelle Studien über die künstliche Abschwächung der Milzbrandbacillen und Milzbrandinfection durch Fütterung. *Mittheilungen an der Kaiserlichen Gesundheitsamte* 2:147–181. *Gesammelte Werke 1*, 232–270.

Koch, Robert. Die Aetiologie der Tuberculose. *Mittheilungen an der Kaiserlichen Gesundheitsamte* 2:1–88 Plates I–X (53 figures). *Gesammelte Werke 1*, 467–565, Plates XX–XXIX.

Koch, R. Ueber die Cholerabakterien. *Deutsche medizinische Wochenschrift* 10:725–728. *Gesammelte Werke 2/1*, 61–68.

1885

Koch, Robert. Lecture and Comments of Robert Koch at the "Conferenz zur Erörterung der Cholerafrage (Zweites Jahr)". *Deutsche medizinische Wochenschrift* 11: Special issue number 37 A, 1–60. *Gesammelte Werke 2/1*, 69–166.

1886

Koch, Robert und G. Gaffky. Versuche über die Desinfection des Kiel-oder Bilgerraumes von Schiffen. *Arbeiten aus dem Kaiserlichen Gesundheitsamte* 1:199–221. *Gesammelte Werke 1*, 380–402.

1887

Gaffky, G. und R. Koch. Bericht über die Thätigkeit der zur Erforschung der Cholera im Jahre 1883 nach Egypten und Indien entsandten Kommission unter Mitwirkung von Dr. Robert Koch bearbeitet von Dr. Georg Gaffky. *Arbeiten aus dem Kaiserlichen Gesundheitsamte* III:1–272, Appendices 1–87, 30 Plates, figures in text. Also published in book form by Springer, Berlin, 1887.

Koch, Robert. *Bericht über die im hygienischen Laboratorium der Universität Berlin ausgeführten Untersuchungen des Berliner Leitungswassers in der Zeit vom 1. Juni 1885 bis 1. April 1886.* Leipzig, Thieme, 1887. *Gesammelte Werke 1*, 410–423.

Koch, Robert. Über die Pasteur'schen Milzbrandimpfungen. Letter published in *Deutsche medizinische Wochenschrift* 13:722. *Gesammelte Werke 1*, 271–273 sowie. Also published in French: De la vaccination charbonneuse. *Semaine médicin* 1887, 305.

1888

Koch, Robert. *Die Bekämpfung der Infektionskrankheiten, insbesondere der Kriegsseuchen.* Berlin, August Hirschwald Verlag, 1888. *Gesammelte Werke 2/1*, 276–289.

1890

Koch, Robert. Ueber bakteriologische Forschung. *Verhandlungen X International Medicin Congress Berlin 1890,* August Hirschwald Verlag, Berlin, Volume I,

35–47. *Gesammelte Werke 1*, 650–660. Review of the lecture: *Deutsche medizinische Wochenschrift* 16:756–757.

Koch, Robert. Weitere Mittheilungen über ein Heilmittel gegen Tuberculose. *Deutsche medizinische Wochenschrift* 16:1029–1032. *Gesammelte Werke 1*, 661–668.

1891

Koch, Robert. Fortsetzung der Mittheilungen über ein Heilmittel gegen Tuberculose. *Deutsche medizinische Wochenschrift* 17:100–102. *Münchner medizinische Wochenschrift* 38:57–58. *Gesammelte Werke 1*, 669–672.

Koch, Robert. Weitere Mittheilung über das Tuberculin. *Deutsche medizinische Wochenschrift* 17:1189–1192. *Münchner medizinische Wochenschrift* 38:769–772 (here with the title: Weitere Mittheilungen über das Tuberculin). *Gesammelte Werke 1*, 673–682.

1893

Koch, Robert. Ueber den augenblicklichen Stand der bakteriologischen Choleradiagnose. *Zeitschrift für Hygiene und Infectionskrankheiten* 14:319–338. *Gesammelte Werke 2/1*, 167–180.

Koch, Robert. Wasserfiltration und Cholera. *Zeitschrift für Hygiene und Infectionskrankheiten* 14:393–426. *Gesammelte Werke 2/1*, 183–206.

Koch, Robert. Entgegnung auf den Vortrag des Herrn Professor Dr. M. Schottelius "Zum mikroskopischen Nachweis von Cholerabacillin in Dejectionen". *Deutsche medizinische Wochenschrift* 19:739. *Gesammelte Werke 2/1*, 181–182.

1896

Koch, Robert and J. Petruschky. Beobachtungen über Erysipel-Impfungen am Menschen. *Zeitschrift für Hygiene und Infectionskrankheiten* 23:477–489. *Gesammelte Werke 2/1*, 267–275.

1897

Koch, Robert. Berichte des Herrn Prof. Dr. Koch über seine in Kimberley gemachten Versuche bezüglich Bekämpfung der Rinderpest. *Zentralblatt für Bakteriologie I. Abteilung. Originale.* 21:526–537. *Gesammelte Werke 2/2*, 690–704.

Koch, Robert. Prof. Robert Koch's Berichte über seine in Kimberley ausgeführten Experimentalstudien zur Bekämpfung der Rinderpest. *Deutsche medizinische Wochenschrift* 23:225–227 and 241–243. *Gesammelte Werke 2/2*, 690–691.

Koch, Robert. Ueber neue Tuberculinpräparate. *Deutsche medizinische Wochenschrift* 23:209–213. *Gesammelte Werke 1*, 683–692.

Koch, Robert. Aerztliche Beobachtungen in den Tropen. *Verhandlungen Deutsche Kolonial-Gesellschaft*, Volume 7, 1897/98. *Gesammelte Werke 2/1*, 326–343.

1898

Koch, Robert. Die Lepraerkrankungen im Kreise Memel. *Klinisches Jahrbuch* 6:239–253. *Gesammelte Werke 2/1*, 670–680.

Koch, Robert. Berichte über die Ergebnisse der Expedition des Geheimen Medizinalrathes Prof. Dr. Koch im Schutzgebiete von Deutsch-Afrika. *Zentralblatt für Bakteriologie I. Abteilung. Originale* 24:200–204. *Gesammelte Werke 2/2*, 733–734 und 731–733.

Koch, Robert. Ueber die Verbreitung der Bubonenpest. *Deutsche medizinische Wochenschrift* 24:437–439. *Gesammelte Werke 2/1*, 647–652.

Koch, Robert. *Reiseberichte über Rinderpest, Bubonenpest in Indien und Afrika, Tsetse—oder Surrakrankheit, Texasfieber, tropische Malariafieber, Schwarzwasserfieber.* Springer, Berlin 1898. *Gesammelte Werke 2/2*, 688–742 (3 figures).

Koch, Robert. Berichte des Geheimen Medizinalrathes Professor Dr. R. Koch über die Ergebnisse seiner Forschungen in Deutsch-Ostafrika. I. Die Malaria in Deutsch-Ostafrika. II. Das Schwarzwasserfieber. *Arbeiten aus dem Kaiserlichen Gesundheitsamte* 14:292–308. *Gesammelte Werke 2/1*, 307–325, 1 Plate (XV), 2 figures.

1899

Koch, R. Ueber Schwarzwasserfieber (Hämoglobinurie). *Zeitschrift für Hygiene und Infectionskrankheiten* 30:295–327. *Gesammelte Werke 2/1*, 348–370.

Koch, Robert. Ueber die Entwicklung der Malariaparasiten. *Zeitschrift für Hygiene und Infectionskrankheiten* 32:1–24, Plates I–IV. *Gesammelte Werke 2/1*, 371–387, Plates XXX–XXXIII.

Koch, Robert. Ergebnisse der wissenschaftlichen Expedition des Geheimen Medizinalrathes Prof. Dr. Koch nach Italien zur Erforschung der Malaria. Vom Kaiserlichen Gesundheitsamte zur Verfügung gestellt. *Deutsche medizinische Wochenschrift* 25:69–70. *Gesammelte Werke 2/1*, 244–347.

Koch, Robert. Erster Bericht über die Thätigkeit der Malariaexpedition. Aufenthalt in Grosseto vom 25. April bis 1. August 1899. Von der Kolonialabtheilung des Auswärtigen Amtes zur Veröffentlichung übergeben. *Deutsche medizinische Wochenschrift* 25:601–604. *Gesammelte Werke 2/1*, 389–396.

1900

Koch, Robert. Zweiter Bericht über die Thätigkeit der Malariaexpedition. Aufenthalt in Niederländisch-Indien vom 21. September bis 12. Dezember 1899. *Deutsche medizinische Wochenschrift* 26:88–90. *Gesammelte Werke 2/1*, 397–403.

Koch, Robert. Dritter Bericht über die Thätigkeit der Malariaexpedition. Untersuchungen in Deutsch-Neu-Guinea während der Monate Januar und Februar 1900. *Deutsche medizinische Wochenschrift* 26:281–284. *Gesammelte Werke 2/1*, 404–411.

Koch, Robert. Vierter Bericht über die Thätigkeit der Malariaexpedition, die

Monate März und April 1900 umfassend. *Deutsche medizinische Wochenschrift* 26:397–398. *Gesammelte Werke 2/1*, 412–415.

Koch, Robert. Fünfter Bericht über die Thätigkeit der Malariaexpedition. Untersuchungen in Neu-Guinea während der Zeit vom 28. April bis zum 15. Juni 1900. *Deutsche medizinische Wochenschrift* 26:541–542. *Gesammelte Werke 2/1*, 416–417.

Koch, Robert. Schlussbericht über die Thätigkeit der Malariaexpedition des Geheimen Medizinalrathes Prof. Dr. Koch. *Deutsche medizinische Wochenschrift* 26:733–734. *Gesammelte Werke 2/1*, 418–419.

Koch, Robert. Zusammenfassende Darstellung der Ergebnisse der Malariaexpedition. *Deutsche medizinische Wochenschrift* 26:781–783 and 801–805. *Gesammelte Werke 2/1*, 420–434.

Koch, Robert. Ergebnisse der vom Deutschen Reiche ausgesandten Malariaexpedition. *Verhandlungen Deutsche Kolonial-Gesellschaft*, Part 1 (1900–1901). *Gesammelte Werke 2/1*, 435–447.

1901

Koch, Robert. Die Bekämpfung der Tuberculose unter Berücksichtigung der Erfahrungen, welche bei der erfolgreichen Bekämpfung anderer Infektionskrankheiten gemacht sind. *Deutsche medizinische Wochenschrift* 27: 549–554. *Gesammelte Werke 1*, 566–567.

Koch, Robert. Ueber die Agglutination der Tuberkelbacillen und über die Verwerthung dieser Agglutination. *Deutsche medizinische Wochenschrift* 27:829–834. *Gesammelte Werke 1*, 694–706.

Koch, Robert. Address on Malaria to the Congress at Eastbourne. *Journal of State Medicine*, Number 10. *Gesammelte Werke 2/1*, 448–455.

Koch, Robert. Ein Versuch zur Immunisirung von Rindern gegen Tsetsekrankheit (Surra). *Deutsche Kolonialblatt* 12: Supplement to Number 1–4. *Gesammelte Werke 2/2*, 743–747.

1902

Koch, Robert. Übertragbarkeit der Rindertuberculose auf den Menschen. Lecture given to the International Tuberculosis Conference in Berlin. *Deutsche medizinische Wochenschrift* 28:857–862. *Gesammelte Werke 1*, 578–590.

Koch, Robert. Framboesia tropica und Tinea imbricata. *Archiv für dermatologie und Syphilis* 59:3–8. *Gesammelte Werke 2/2*, 681–687 (5 figures).

Koch, R., und W. Schütz. Menschliche Tuberculose und Rindertuberculose (Perlsucht). *Archiv für wissenschaftliche und praktische Tierheilkunde* 28:169–196. (Not in *Gesammelte Werke*.)

1903

Koch, Robert. Die Bekämpfung der Malaria. *Zeitschrift für Hygiene und Infectionskrankheiten* 43:1–4. *Gesammelte Werke 2/1*, 456–458.

Koch, Robert. Die Bekämpfung des Typhus. *Veröffentlichungen auf dem Gebiete des Militärischen Sanitäts-Wesens 1903*, Volume 21, Hirschwald, Berlin. *Gesammelte Werke 2/1*, 296–305.

1904

Koch, Robert. Ueber die Trypanosomenkrankheiten. *Deutsche medizinische Wochenschrift* 30:1705–1711. *Gesammelte Werke 2/1*, 459–472 (5 figures).

Koch, Robert. Untersuchungen über Schutzimpfungen gegen Horse-Sickness (Pferdesterbe). *Deutsche Kolonialblatt* 15:420–424. *Berliner tierärztliche Wochenschrift* 20:546–547. *Gesammelte Werke 2/2*, 774–787.

Koch, Robert. Fourth Report on African Coast Fever. Presented to the Legislative Council, 1904. *Gesammelte Werke 2/2*, 787–798.

Koch, Robert. Vorläufiger Bericht über das Rhodesische Rotwasser-oder "Afrikanische Küstenfieber". *Archiv für wissenschaftliche und praktische Tierheilkunde* 30:281–295. *Gesammelte Werke 2/2*, 748–757.

Koch, Robert. Zweiter Bericht über das Rhodesische Rotwasser-oder "Afrikanische Küstenfieber". Translated from English. *Archiv für wissenschaftliche und praktische Tierheilkunde* 30:295–304. *Gesammelte Werke 2/2*, 757–763.

Koch, Robert. Dritter Bericht über das Rhodesische Rotwasser-oder "Afrikanische Küstenfieber". Translated from English. *Archiv für wissenschaftliche und praktische Tierheilkunde* 30:305–319. *Gesammelte Werke 2/2*, 764–773.

Koch, Robert. Vierter Bericht über das Rhodesische Rotwasser-oder "Afrikanische Küstenfieber". Translated from English. *Archiv für wissenschaftliche und praktische Tierheilkunde* 30:586–598. Not in *Gesammelte Werke*. Also in English in: *Agricultural Journal of the Cape of Good Hope* 24:549–557.

1905

Koch, Robert, W. Schütz, F. Neufeld, and H. Miessner. Ueber die Immunisirung von Rindern gegen Tuberculose. *Zeitschrift für Hygiene und Infectionskrankheiten* 51:958–962. *Gesammelte Werke 1*, 591–611.

Koch, Robert. Ueber die Unterscheidung der Trypanosomenarten. *Sitzungsbericht der Königlich Preussischen Akademie der Wissenschaften, Physik und Medizin, Klasse XLVI*, 958–962. *Gesammelte Werke 2/1*, 473–476.

Koch, Robert. Vorläufige Mitteilungen über die Ergebnisse einer Forschungsreise nach Ostafrika. *Deutsche medizinische Wochenschrift* 31:1865–1869. *Gesammelte Werke 2/1*, 477–486 (24 figures).

Koch, Robert. Ueber den derzeitigen Stand der Tuberculosebekämpfung. Nobel Prize address delivered 12 December 1905 in Stockholm. *Deutsche medizinische Wochenschrift* 32: (1906):89–92. *Gesammelte Werke 1*, 612–619.

Koch, Robert. Zwei Berichte über Pferdesterbe. *Archiv für wissenschaftliche und praktische Tierheilkunde* 31:330–334. Not in *Gesammelte Werke*.

1906

Koch, Robert. Beiträge zur Entwicklungsgeschichte der Piroplasmen. *Zeitschrift für Hygiene und Infectionskrankheiten* 54:1–9, Plates I–III. *Gesammelte Werke* 2/1, 487–492, Plates XXXIV–XXXVI.

Koch, Robert. Ueber afrikanischen Recurrens. *Berliner klinische Wochenschrift* 43:185–194. *Gesammelte Werke* 2/1, 493–508 (10 figures).

Koch, Robert. Ueber den bisherigen Verlauf der deutschen Expedition zur Erforschung der Schlafkrankheit in Ostafrika. *Deutsche medizinische Wochenschrift* 32: Special supplement to Number 51, I–VIII. *Gesammelte Werke* 2/1, 509–524.

1907

Koch, Robert. Bericht über die Tätigkeit der deutschen Expedition zur Erforschung der Schlafkrankheit bis zum 25. November 1906 (aus Sese bei Entebbe an den Staatssekretär des Innern). *Deutsche medizinische Wochenschrift* 53:49–51. *Gesammelte Werke* 2/1, 525–530.

Koch, Robert. Bericht des Herrn Geheimen Medizinal-Rathes Prof. Robert Koch von der deutschen Expedition zur Erforschung der Schlafkrankheit. *Deutsche medizinische Wochenschrift* 33:1462–1463. *Gesammelte Werke* 2/1, 531.

Koch, Robert. Schlussbericht über die Tätigkeit der Deutschen Expedition zur Erforschung der Schlafkrankheit. *Deutsche medizinische Wochenschrift* 33:1889–1895. *Gesammelte Werke* 2/1, 534–546.

1908

Koch, Robert. *Ueber meine Schlafkrankheitsexpedition.* Reimer, Berlin 1908. *Gesammelte Werke* 2/1, 563–581 (22 figures).

Koch, Robert. Anthropologische Beobachtungen gelegentlich einer Expedition an den Victoria-Nyanza. *Zeitschrift für Ethnologie* 40:449–470. *Gesammelte Werke* 2/1, 547–562 (20 figures).

Koch, Robert. Das Verhältnis zwischen Menschen—und Rindertuberculose (Translated from English). *Berliner klinische Wochenschrift* 45:2001–2003. *Gesammelte Werke 1*, 624–635. Lecture given at the International Tuberkulosis Congress in Washington, 30 September 1908 and in a special session on 2 October 1908.

1909

Koch, Robert. Antrittsrede des Herrn Koch. Antrittsrede in der Preussischen Akademie der Wissenschaften am 1. Juli 1909. *Deutsche medizinische Wochenschrift* 35:1278–1279. Abhandlungen der Deutschen Akademie der Wissenschaften zu Berlin, Klasse für Medizin, Jahrgang 1960, Nr. 6, 38–41. Akademie Verlag 1961, Berlin. *Gesammelte Werke 1*, 1–4.

1910

Koch, Robert. Epidemiologie der Tuberculose. *Zeitschrift für Hygiene und Infectionskrankheiten* 67:1–18. *Gesammelte Werke 1*, 636–649. Speech published posthumously. Given to the Akademie der Wissenschaften zu Berlin on 7. April 1910.

1911

Koch, R., M. Beck, and F. Kleine. Bericht über die Tätigkeit der zur Erforschung der Schlafkrankheit im Jahre 1906/07 nach Ostafrika entsandten Kommission. *Arbeiten aus dem Kaiserlichen Gesundheitsamte* 31:1–320 (100 figures, 5 Plates). *Gesammelte Werke 2/1*, 582–645 (34 figures, Plates XXXIX–XL).

The following book is a compilation in English of Robert Koch's key papers: Carter, K. Codell (editor) 1987. *Essays of Robert Koch*. Greenwood Press, Westport, Connecticut. 188 pages.

II. Biographies of Robert Koch*

Becher, W. *Robert Koch—Eine bibliographische Studie*. H. Cornitzers Verlag, Berlin. 1890. 54 pages. This was the first biography of Koch, published when he was only 47 years old.

Bochalli, R. *Robert Koch—Der Schöpfer der modernen Bakteriologie*. Wissenschaftliche Verlagsgesellschaft. Stuttgart 1954, 184 pages, 12 figures. Brief and with some errors, but still in print.

Genschorek, W. *Robert Koch—Leben, Zeit, Werk*. 3rd edition. 1979, S. Hirzel Verlag Leipzig. 206 pages, 90 figures. Somewhat more detailed than Bochalli; still in print.

*In addition to the biographies listed here, there have been several German-language novels based on Koch's life.

Heymann, Bruno. *Robert Koch. 1. Teil 1843–1882.* Leipzig, Akademische Verlagsgesellschaft. 1932. 352 pages. Frontispiece and 15 figures. The best biography of Koch, but only the first part was published.

Kirchner, M. *Robert Koch.* Rikola Verlag Wien, Berlin, Leipzig, München 1924. An early biography; by one of Koch's close associates.

Kitasato, S. *The Battle against Tuberculosis. Robert Koch's Life Work* (Japanese Text). Tokyo, 1913. (Cited by Fox, H.: Baron Shibasaburo Kitasato, *Annals of Medical History 6* (1934), 491–499.)

Lagrange, E. *Robert Koch—Sa Vie et son Oeuvre.* En Dépòt chez M. Legrand libraire Paris à 1 'Edition Universelle Bruxelles 1938. 90 pages, 10 figures.

Löbel, J. *Robert Koch. Geschichte eines Glücklichen.* Zürich 1935, 320 pages.

Metschnikov, E. *Pasteur, Lister, Koch—The Founders of Modern Medicine.* Moscow 1915 (Russian). An English edition was published by Walden Publications, New York, 1939. It included translation of Koch's *Etiology of Wound Infections.* 387 pages.

Möllers, Bernhard. *Robert Koch—Persönlichkeit und Lebenswerk 1843–1910.* Schmorl und Seefeld Nachf. Hannover 1951, 756 pages, 76 figures. Detailed biography by one of Koch's last students. Extensive publication of Koch's letters. (Publication delayed because of World War II.)

Wezel, K. *Robert Koch. Eine biographische Studie.* Verlag von August Hirschwald, Berlin 1912, 148 pages.

Notes

Chapter 2 *Koch's Early Years*

1. Koch, Robert. 1909. *Deutsche Medizinische Wochenschrift* 35:1278–1279.

2. Koch eventually became quite proficient in English, but he was never fluent in French, and one of the sources of his later conflict with Louis Pasteur (Chapter 16) was Koch's unfamiliarity with spoken French.

3. Möllers, Bernhard. 1950. *Robert Koch. Persönlichkeit und Lebenswerk. 1843–1910.* Schmorl and Von Seefeld, Hannover, page 23.

4. Henle's presumed influence on Koch's later career is widely quoted. In his book published in 1840, Henle hypothesized that contagious diseases were caused by specific living pathogens, the so-called "contagium animatum", and outlined the three-fold proof of an infectious agent: microscopy, culture, animal experimentation. However, although Koch himself, in an address to the Royal Prussian Academy of Sciences, mentioned Henle as one of his teachers, he did not emphasize Henle's influence on his career. The key passage in Koch's speech is as follows: "If I now consider the academic subjects which influenced my scientific development and especially my relationship to bacteriology, I must first indicate that I received from the University no direct influence for my scientific direction, because bacteriology did not exist in the University. Yet I would like to give special thanks to several of my teachers: the anatomist Henle, the clinician Hasse, and especially the physiologist Meissner, who awoke in me a feeling for scientific research." However, Henle's influence on experimental medicine in mid-nineteenth century Germany was so vast that Koch may have owed a greater debt to Henle's teachings than he perhaps realized.

5. Address to the Royal Prussian Academy of Sciences. 1 July 1909. *Gesammelte Werke*, Volume I, pages 1–4. Also *Deutsche Medizinische Wochenschrift* 35:1278–1279. Koch was the first bacteriologist admitted into the Academy of Sciences.

6. Although his teachers certainly contributed, the person most responsible for Koch's career and success was the botanist Ferdinand Cohn (1828–1898; see Chapter 6).

7. Koch's paper, with the colored drawings, is reproduced in his Collected Works: *Gesammelte Werke*, 2 (2): 806–813.

8. Möllers, page 33.

9. Koch, R. 1865. Über das Entstehen der Bernsteinsüre im menschlichen Organismus. *Zeitschrift für rationelle Medizin*, 24: 264–274.

Chapter 3   The Young Doctor and Husband

1. Möllers, op.cit., page 65.

2. Möllers, op. cit., page 65.

3. Möllers, op. cit., page 67.

Chapter 4   Koch in Wollstein

1. Pfuhl, Eduard. 1912. Robert Koch's Entwicklung zum bahnbrechenden Forscher. *Deutsche Medizinische Wochenschrift* 37:1101–1102, 1148–1150, 1195–1197.

2. Its name was changed to Wolsztyn after it became part of Poland following Germany's defeat in World War I. Koch's house is still standing (see Figure 4.1).

3. Heymann, Bruno. 1932. *Robert Koch. I. Teil. 1843–1882.* Akademische Verlagsgesellschaft, Leipzig, pages 109–111. The planned second volume never appeared.

4. Pfuhl, Gertrud. 1940. Robert Koch in Wollstein. Erinnerungen seiner Tochter Gertrud Pfuhl. *Deutsche Medizinische Wochenschrift* 66:355–357.

5. Möllers, op. cit., page 97.

6. Pfuhl, loc. cit., 1912.

7. Möllers, op. cit., page 97.

8. Pfuhl, loc. cit., 1912.

Chapter 5   The Work on Anthrax

1. Doyle, Arthur Conan. 1890. Character sketch of Dr. Robert Koch. *Review of Reviews* December 1890, page 552.

2. Henle, J. 1840. On Miasma and Contagion, in *Pathologische Untersuchungen.* August Hirschwald Verlag, Berlin.

3. Klebs was born in Königsberg in 1834, studied there and in Würzburg and Jena before going on to Berlin where he received his degree. After brief medical practice in Königsberg he turned to academic work in pathology and physiology and returned to Berlin to study under Virchow. His first academic position was in Bern, Switzerland. Never remaining long in one place, he served successively in Würzburg, Prague, Zurich, Karlsruhe, Strassburg, Asheville (North Carolina), Chicago (Rush Medical College), Hannover, Berlin, Lausanne, and Bern, where he died in 1913 after a long and active life. See Carter, K.C., 1985, (Koch's Postulates in relation to the work of Jacob Henle and Edwin Klebs, *Medical History* 29:353–374.) for a discussion of the work of Klebs.

4. Klebs was also honored by having an organism, *Klebsiella pneumoniae*, named for him, although this is a misnomer since the organism was not discovered by Klebs but by Carl Friedländer (see Mochmann and Köhler, op. cit., pages 219–222).

5. Klebs, E. 1877. Über die Umgestaltung der medicinischen Anschauungen in den letzten drei Jahrzehnten. Lecture given to the *50 Versammlung Deutscher Naturforscher und Ärzte in München*, page 47.

6. Heymann, op. cit., page 123.

7. Davaine, C.J. 1863. Recherches sur les infusoires du sang dans la maladie connue sur le nome de sang de rate. *Comptes rendu de l'academie des sciences, Paris* Volume LVII.

8. Heymann, op. cit., page 123.

9. Koch's notebook, as quoted by Heymann, op. cit., page 140.

10. Julius Cohnheim, the well-known Breslau pathologist and later one of Koch's most enthusiastic champions, had pioneered the use of the rabbit cornea for studying infectious disease.

11. Koch, Robert. 1876. Die Ätiologie der Milzbrandkrankheit, begründet auf die Entwicklungsgeschichte des Bacillus Anthracis. *Beiträge zur Biologie der Pflanzen* 2:277–310.

12. Koch, 1876, loc. cit.

13. Koch, 1876, loc. cit.

14. Pfuhl, 1912, op. cit.

15. The story of how Koch got white mice is described by Möllers (op. cit., page 461). An assistant of Ferdinand Cohn, Eduard Eidam, studied with Koch in Wollstein in 1876. Eidam was accompanied by his foster father, a kindly old gentleman who loved children and who made friends during his stay with Koch's daughter Gertrud. Upon returning to Berlin, the old gentleman sent the daughter some pet white mice in a special "mouse house" (a cage in the form of a little house, with floors, tiny rooms, stairways and windows). The mice multiplied rapidly and Koch began to use the "excess" mice for his experiments. It was in this way that the white mouse was first introduced by Koch as an experimental animal.

16. Koch, 1876, loc. cit.

17. Heymann, op. cit., page 147.

Chapter 6 *Koch and Cohn*

1. Heymann, op. cit., page 148.

2. Part of Poland since the end of World War I and now known as Wroclaw.

3. The background material on Ferdinand Cohn is taken from: Pauline Cohn, *Ferdinand Cohn. Blätter der Erinnerung.* J.U. Kern's Verlag, Breslau, 1901 and Rosen, F., 1922, in Andreae, F., M. Hippe, O. Schwarzer, and H. Wendt, (editors) *Schlesier des 19.Jahrhunderts,* Korn Verlag, Breslau, pages 167–173.

4. Cohn, F. 1872. *Ueber Bacterien, die kleinsten lebenden Wesen,* C.G. Lüderitz'sche Verlagsbuchhandlung, Berlin. An English translation of this book was published: Dolley, Charles S. 1881. Bacteria: the smallest of living organisms. Rochester, N.Y. Reprinted in 1939 by the Johns Hopkins Press, Baltimore.

5. Pauline Cohn, loc. cit.

6. Heymann, op. cit., page 178.

7. The Koch visit is recorded in the Institute log book in Cohn's own hand. These accounts were published by Heymann, pages 150–154.

8. The information on Cohnheim is taken from the introduction to Cohnheim, Julius F. 1885. *Gesammelte Abhandlungen* edited by E. Wagner. A. Hirschwald Verlag, Berlin.

9. Wagner, loc. cit.

10. Koch, 1876, op. cit.

11. Cohn, F. 1876. Untersuchungen über Bacterien. IV. Beiträge zur Biologie der Bacillen. *Beiträge zur Biologie der Pflanzen,* 2: 249–276.

12. Koch, 1876, op. cit.

13. Tyndall, John. 1877. Further researches on the deportment and vital persistence of putrefactive and infective organisms from a physical point of view. *Philosophical Transactions of the Royal Society of London* 167: 149–206. Tyndall's fractional sterilization technique is sometimes called *Tyndallization.*

14. Heymann, op. cit., page 159.

15. Heymann, op. cit., page 172.

16. Heymann, op. cit., pages 173–175.

Chapter 7 *The Microscope Revolution*

1. Quoted from Cohn (1872), page 15.

2. Reichardt, Oscar and C. Stürenburg. 1868. *Lehrbuch der Mikroskopischen Photographie.* Verlag von Quandt & Händel, Leipzig.

3. Heymann, op. cit., page 167.

4. Heymann, loc. cit.

5. Koch, R. 1877. Verfahren zur Untersuchung, zum Conservieren und Photographiren der Bakterien. *Beiträge zur Biologie der Pflanzen,* 2:399–434.

6. Quoted from Heymann, page 171. Written 14 October 1876.

7. Heymann, loc. cit.

8. Reichert and Stürenburg, loc. cit.

9. Gerlach, J. 1863. *Die Photographie als Hülfsmittel Mikroskopischer Forschung.* Verlag von Wilhelm Engelmann, Leipzig.

10. Sources of photographic collodion in Koch's time were E. Liesegang of Elberfeld and Ferdinand Beyrich of Berlin.

11. Gerlach, loc. cit.

12. Heymann, op. cit., page 173–175.

13. Heymann, op. cit., page 176.

14. Heymann, op. cit., page 179. Note that this letter was written on Christmas day, 1876!

15. Koch, 1877, loc. cit.

16. Weigert, C. 1878. Bismarckbraun als Färbemittel. *Archiv für mikroskopische Anatomie* 15: 258–260.

17. Salomonsen, C.J. 1914. Lebenserinnerungen aus dem Breslauer Sommersemester 1877. *Berl. klin. Wochenschr.* 51:485–490. Another visitor at this same time was William Henry Welch, from America, soon to become Professor of Pathology at the Johns Hopkins University. Welch became one of Koch's main supporters in the United States (see Chapter 21). Because of Welch's enormous influence on American medicine, Koch's work and methods became widely known at an early stage in the United States.

18. Koch later found out that he was overly optimistic about photomicrographs. See the preface to his 1878 book (discussed in Chapter 8).

19. Salomonsen, loc. cit. Salomonsen also took Koch's cholera course in Berlin in the mid-1880's.

20. Heymann, op. cit., page 186.

21. Heymann, op. cit., page 195.

22. Heymann, op. cit., page 201.

23. Heymann, op. cit., page 224.

24. Bradbury, S. 1967. *The Evolution of the Microscope.* Pergamon Press, Oxford.

25. Abbe, Ernst. 1879. On Stephenson's system of homogeneous immersion for microscope objectives. *Transactions of the Royal Microscopical Society* 2: 256–265.

26. Koch, 1877, loc. cit.

Chapter 8 *Studies on Wound Infections*

1. Heymann, op. cit., page 257.

2. Heymann, op. cit., page 197. Koch was later to write very critically of Pasteur's work on anthrax, initiating a running feud with the eminent French scientist that was to last for many years (Chapter 16).

3. Welch, W.H. Collected Works. See also Chapter 21.

4. Heymann, op. cit., page 217–218.

5. Another powerful opponent of distinct bacterial species was the Viennese surgeon Theodor Billroth, who believed that all bacteria were but stages in the life cycle of an alga which he called Coccobacteria septica. Billroth published his idea in a lengthy book, a copy of which he sent to Cohn. Billroth's work was technically flawed and his book was full of confusing fantasies, but Cohn dealt with them temperately, gently pointing out Billroth's errors. Billroth, T. 1874. *Untersuchungen über die Vegetationsformen von Coccobacteria septica und den Antheil, welchen sie an der Entstehung und Verbreitung der accidentellen Wundkrankheiten haben. Versuch einer wissenschaftliche Kritik der verschiedenen Methoden der antiseptischen Wundbehandlung.* Reimer, Berlin.

6. Nägeli, Carl von. 1877. *Die niederen Pilze in ihren Beziehungen zu den Infectionskrankheiten und der Gesundheitspflege.* R. Oldenbourg, München.

7. Nägeli, loc. cit.

8. Letter written 11 November 1877. Quoted from Heymann, page 218. Koch also wrote a lengthy and highly critical review of Nägeli's book, and of another book published at the same time by one of Nägeli's supporters, Hans Buchner (*Die Nägeli'sche Theorie der Infectionskrankheiten in ihren Beziehungen zur medicinischen Erfahrung,* 1877, Leipzig, 112 pp).

9. The insistence by Koch and Cohn on the immutability of bacterial species may have been the appropriate posture in 1877, but it subsequently caused a serious delay in the development of the field of bacterial genetics. This was most clearly stated in Hadley's early important paper on bacterial variation: "The doctrine of monomorphism has descended to us from the early conceptions of the nature of bacteria maintained by Cohn, Koch, and others of the early school. Under its influence . . . there were set up strict notions of 'normal' bacterial cell types, 'normal' colony forms, and 'normal' cultures. Whatever departed from the expected normality was at once relegated to the field of contaminations; or to the weird category of 'involution forms', 'degeneration forms,' or pathological elements possessing neither viability, interest, nor significance. . . . Although some early opposition to these views arose, it was probably unfortunate for the beginnings of the science that the first attempts toward modification were made by such extremists as Nägeli and his associates . . . The extreme plurimorphism which they so eagerly championed through years of bitter controversy was too radical to be accepted graciously as an antidote to strict monomorphism. . . . The interpretation of the Berlin school triumphed— and to such an extent as to become later the dogma of 'normal' colony and culture types that has endured, with hardly a respite, even to the present day." Hadley, Philip. 1927. The dissociative aspects of bacterial behavior. pp. 84–85 in E.O. Jordan and I.S. Falk (editors), *The Newer Knowledge of Bacteriology and Immunology,* University of Chicago Press, Chicago.

10. Koch, R. 1878. *Untersuchungen über die Aetiologie der Wundinfectionskrankheiten.* Vogel, Leipzig.

11. Klebs, Edwin. 1872. *Beiträge zur pathologischen Anatomie der Schusswunden.* Vogel, Leipzig.

12. See especially Billroth's book, loc. cit.

13. Bulloch, W. 1938. *A History of Bacteriology.* Oxford University Press, Oxford.

14. Bulloch, loc. cit., page 147.

15. Koch, 1878, loc. cit., page 15.

16. Weigert reported the value of aniline staining at a meeting in Breslau: *Bericht über die Sitztungen der Schlesischen Gesellschaft für Vaterländische Cultur,* December 10, 1878. It was published in more detail in *Archiv für mikroskopische Anatomie* 15: 258–260. 1878.

17. Heymann, op. cit., page 246.

18. Quoted from the paper Koch read at Kassel: Koch, R. 1878. Neue Untersuchungen über die Mikroorganismen bei infectiösen Wundkrankheiten. *Mittheilungen 51 deutschen Naturforscherversammlung zu Cassel. Deutsche medizinische Wochenschrift* 26 October 1878, page 531.

19. Quoted from Translator's Preface to Koch, R. 1880. *Investigations into the Etiology of Traumatic Infective Diseases,* translated by W. Watson Cheyne. New Sydenham Society, London.

20. Koch, reference 10, page 68.

21. Koch, loc. cit., page 68–69.

22. Koch, loc. cit., page 71.

23. Ogston was so impressed with Lister's antiseptic surgery that after seeing it demonstrated he went into the operating room of the Aberdeen hospital where he worked and tore down the sign that read: "Prepare to meet thy God"! (Quoted from Mochmann and Köhler, op. cit., page 137.

24. Ogston, Alexander. 1881. Report upon micro-organisms in surgical diseases. *British Medical Journal* March 12, 1881, pages 369–375. The passage quoted is from page 369.

25. Heymann, op. cit., page 257.

26. Möllers, op. cit., pages 110–111. As another indication of Virchow's attitude to bacteria, Carl Salomonsen's experience can be cited: "During my visit to Virchow in Berlin . . . I told him that I was especially involved in bacteriological studies. He made several remarks about the difficulty of pure culture and then added, laughingly: 'Brefeld once told me: In order to be certain what one was doing, one must have a special small laboratory for each species!' " Salomonsen, C.J. 1914. Lebenserinnerungen aus dem Breslauer Sommersemester 1877. *Berl. klin. Wochenschr.* 51:485–490.

27. It has often been said that authority breeds contempt. In his later years, Koch himself acquired the authority of a Virchow, and became equally intransigent about the work of others (see Chapters 21 and 22).

28. Koch, R. 1878. Neue Untersuchungen über die Mikroorganismen bei infectiösen Wundkrankheiten. *Deutsche medizinische Wochenschrift* 4: 531–532.

29. Bulloch, loc. cit., page 152.

Chapter 9 *On to Berlin*

1. Heymann, op. cit., page 289–290.

2. Heymann, op. cit., pages 272–273.

3. Heymann, op. cit., page 289–290.

4. Heymann, op. cit., page 291.

5. Heymann, op. cit., page 291.

Chapter 10 *Koch at the Crossroads*

1. Salomonsen, C.J. 1914. Lebenserinnerungen aus dem Breslauer Sommer—Semester 1877. *Berliner Klinishe Wochenschrift*, page 486.

2. Loeffler, R. 1903. Robert Koch. Zum 60 Geburtstage. *Deutsche medizinische Wochenschrift*, page 938.

3. Loeffler worked in the Imperial Health Office for five years under Koch. In 1888 he was appointed Professor of Hygiene at the University of Greifswald and in 1913 became the third director of the Robert Koch Institute of Infectious Disease in Berlin. He died in 1915.

4. Gaffky left the Imperial Health Office in 1888 to become Professor of Hygiene at the University of Giessen. He and Koch remained close friends and colleagues and through the years carried on an extensive correspondence (quoted in detail in Möllers, op. cit.). Gaffky subsequently became the second director of the Robert Koch Institute of Infectious Disease, assuming this position when Koch retired from the post in 1904. He died in 1918.

5. Salomonsen, loc. cit.

6. Mittheilungen aus dem Kaiserlichen Gesundheitsamte, Volume 1, edited by Dr. Struck, Berlin, 1881, 399 pages and 16 pages of plates.

Chapter 11 *The Plate Technique*

1. Koch, 1881. Zur Untersuchungen von pathogenen Organismen. *Mittheilungen aus dem Kaiserlichen Gesundheitsamte*, page 18.

2. When speaking of cultures, "purity" is more of a concept than an actuality. One can never prove that a culture is definitely pure. All one can do is show that the probability of contamination is very low.

3. Lister, Joseph. 1878. On the lactic fermentation and its bearing on pathology. *Transactions of the Pathological Society of London* 29: 425–467.

4. Lister, 1878, loc. cit.

5. Koch, 1881, loc. cit. pages 1–48. The paper was completed on 10 May 1881, less than a year after Koch came to Berlin.

6. Schroeter, J. 1875. Ueber einige durch Bacterien gebildete Pigmente. *Beiträge zur Biologie der Pflanzen* 1: 109–126.

7. Brefeld, O. Methoden zur Untersuchungen der Pilze. *Verhandlungen der phys.-med. Gesellsch. in Würzburg,* 8: 43–62.

8. Heymann, op. cit., page 304.

9. The cotton plug had been introduced into bacteriological research many years earlier. See Dusch, Theodor F. and H. Schröder. 1854. Ueber Filtration der Luft in Beziehung auf Fäulniss und Gährung. *Ann. der Chem. V. Pharm.* 89:232.

10. We will discuss Koch's conflict with Pasteur in detail in Chapter 16.

11. Koch, 1881, loc. cit., pages 19–21.

12. Neubaur, 1883. XI. Oeffentliches Sanitätswesen. 1. Allgemeine Deutsche Ausstellung auf dem Gebiete der Hygiene und des Rettungswesens. X. Die Demonstrationen im Pavillion des Reichs-Gesundheitsamtes mit besonderer Berücksuchtigung der bei ihnen verwendeten Apparate. XI. This was a general German exhibition on hygiene and public health. The demonstrations were held in the pavillion of the Imperial Health Office, showing examples of the kinds of equipment used. *Deutsche medicinische Wochenschrift* 9: 573–575.

13. Hitchens, A.P. and M.C. Leikind. 1939. The introduction of agar-agar into bacteriology. *Journal of Bacteriology* 37: 485–493. See also v. Gierke, 1935, Zur Einführung des Agar-Agars in die bakteriologische Technik. *Zentralblatt für Bakteriologie, Parasitenkunde und Infektionskrankheiten* 133: 273.

14. Petri, R.J. 1887. Eine kleine Modification des Koch'schen Plattenverfahrens. *Centrallblatt für Bacteriologie und Parasitenkunde* 1: 279–280. It is an accident of history that Petri's name is almost better known than Koch's. I can't resist quoting the following impressions of Petri by a contemporary. When the following writer knew him, Petri was chief of a tuberculosis sanitorium, despite the fact that he was not a physician. "When I first met him Petri was getting on for sixty. He was an authentic Prussian disciplinarian, the strict and rather vain headmaster type of man I have always abominated. His chief anxiety was to maintain discipline not only amongst his staff, but amongst his patients too. On any and every half-way suitable occasion he would appear in the full-dress uniform of a Chief Army Doctor, and the sash round his protuberant belly always reminded me of the equator round a globe. The man wasn't a doctor at all, and over and above that he was falling into premature senility, but in his lucid moments one could learn a thing or two from him where bacteriology and laboratory technique were concerned. He had made quite a name for himself in the scientific world by a minor but brilliant process with which he solved the problem of isolating microscopic individual phenomena from the general convolut of bacteria . . . His 'Petri-dish' was the material key to the subsequent tremendous development of bacteriology . . ." Plesch, John. 1947. *Janos, The Story of a Doctor,* Victor Gollancz Ltd., London, pages 51–52.

15. Hueppe, Ferdinand. 1889. *Die Methoden der Bakterien-Forschung.* C.W. Kreidel's Verlag, Wiesbaden, page 296.

16. Hueppe, loc. cit.

Chapter 12 *Sterilization and Disinfection*

1. Bulloch, William. 1938. *A History of Bacteriology*, Oxford University Press, London, page 235.

2. Koch, Robert. 1881. Ueber Desinfection. *Mittheilungen aus dem kaiserlichen Gesundheitsamt* 1: 234–282.

3. Lister, Joseph. 1884. An address on corrosive sublimate as a surgical dressing. *British Medical Journal* 2: 803.

4. Krönig, B. and Th. Paul. 1897. Die chemischen Grundlagen der Lehre von der Giftwirkung und Desinfection. *Zeitschrift für Hygiene und Infectionskrankheiten*, 25: 1–112.

5. Koch, Robert and Gustav Wolffhügel. 1881. Untersuchungen über die Desinfection mit heisser Luft. *Mittheilungen aus dem kaiserlichen Gesundheitsamt* 1: 301–321.

6. Koch, Robert, Georg Gaffky and Friedrich Loeffler. 1881. Versuche über die Verwerthbarkeit heisser Wasserdämpfe zu Desinfectionszwecken. *Mittheilungen aus dem kaiserlichen Gesundheitsamt* 1: 322–340.

7. In this dry heating procedure the steam itself never came in contact with the organisms, of course, since it was confined to the inside of the copper tubing.

8. Electrical heating, which would be used today, was not, of course, available at this time. Electrification of buildings did not occur until near the end of the 19th century.

Chapter 13 *The London Meeting*

1. Godlee, R.J. 1924. *Lord Lister*. Oxford University Press, Oxford, page 446.

2. Lister, Joseph. 1881. On the relations of minute organisms to unhealthy processes arising in wounds, and to inflammation in general. *Transactions of the International Medical Congress*, Seventh Session, Volume I, pages 311–312.

3. Godlee, loc. cit.

4. Lister, Joseph. 1890. An address on the present position of antiseptic surgery. *British Medical Journal*, 2: 377.

5. Godlee, loc. cit.

Chapter 14 *The Tubercle Bacillus*

1. Loeffler, F. 1907. Zum 25 jährigen Gedenktage der Entdeckung des Tuberkelbacillus. *Deutsche Medizinische Wochenschrift* 33: 449–451.

2. New York Times, May 3, 1882.

3. Dubos, René and Jean Dubos. 1956. *The White Plague*. Little, Brown, Boston.

4. "The captain of all these men of death that came against him to take him away, was the Consumption, for it was that that brought him down to the grave." John Bunyan. 1680. *The Life and Death of Mr. Badman.*

5. Koch, Robert. 1882. Die Aetiologie der Tuberculose. *Berliner Klinische Wochenschrift* 19: 221–230. A useful short discussion of the background for this paper is given by Schadewaldt, H. 1982. Wissenschaftlich Wertarbeit in sechs Monaten. *Mk. Ärztl. Fortb.* 32: 24–33.

6. We can be amazed, even in this age of computerized publication, how *rapidly* papers were published in Koch's time. Working from a handwritten manuscript using handset type (this was before typewriters and linotype), Koch's paper was brought into print in less than three weeks!

7. Dubos and Dubos, loc. cit.

8. Villemin's first paper on the transmissibility of tuberculosis to experimental animals was published in 1865 in *Compt. rend. Acad. d. Sci.* Paris, 61: 1210. In 1868 he published a book on the same subject: *Études sur la tuberculose*. Paris. 594 pages.

9. Loeffler, loc. cit.

10. Ehrlich, Paul. 1882. Modification der von Koch angegebene Methode der Färbung von Tuberkelbacillen. *Deutsche Medizinische Wochenschrift* 8: 269–270.

11. Koch, 1882, loc. cit. Interestingly, Koch recognized even early in his work that not only the tubercle bacillus, but also the bacillus associated with leprosy, had these unusual staining properties. The importance of these staining properties as an indication of a *specific* organism as causal agent of tuberculosis was obvious.

12. Koch, 1882. loc. cit.

13. This section is based on Loeffler's article, loc. cit. and Heymann's book, op. cit., pages 310–330.

14. The room at the Humboldt University (East Berlin) where Koch gave this famous lecture is still extant (see Figure 14.3) and can be visited. A plaque on the wall commemorates the important occasion.

15. Loeffler, loc. cit.

16. Loeffler, loc. cit.

17. Möllers, op. cit., page 133.

18. Koch, Robert. 1882. Die Aetiologie der Tuberculose. *Berliner klinischer Wochenschrift* 19: 221–230.

19. London Times, April 22, 1882.

20. New York Times, May 3, 1882.

21. Animal rights: some issues never go away!

22. London Times, April 22, 1882.

23. New York Tribune, May 3, 1882.

24. Ehrlich, loc. cit.

25. Ziehl, Franz. 1882. Zur Färbung des Tuberkelbacillus. *Deutsche Medizinische Wochenschrift* 8: 451.

26. Neelsen, Friederich. 1883. Ein casuistischer Beitrag zur Lehre von der Tuberkulose. *Centralblatt Medizinische Wissenschaften* 21: 497–501.

27. Gram, Christian. 1884. Über die isolirte Färbung der Schizomyceten in Schnitt-und Trockenpräparaten. *Fortschritte der Medizin* 2: 185–189.

28. Baumgarten, Paul. 1882. Tuberkelbakterien. *Centralblatt Medizinische Wissenschaft* 20: 257–259.

29. Koch, Robert. 1883. Kritische Besprechungen der gegen die Bedeutung der Tuberkelbazillen gerichteten Publikationen. *Deutsche Medizinische Wochenschrift* 9: 137–141.

30. Formad, H.F. 1884. The bacillus tuberculosis and the aetiology of tuberculosis. Is consumption contagious? *Journal of the American Medical Association* 2: 141–151 (see this paper also for references to earlier papers). This controversy has also been summarized by Landis, H.R.M. 1932. The reception of Koch's discovery in the United States. *Annals of Medical History* 4: 531–537.

31. Spina, A. 1883. *Studien über Tuberkulose.* Vienna.

32. Koch, 1883. loc. cit.

33. Shakespeare, Edward O. 1884. The bacillus theory of tuberculosis. *New York Medical Journal* 39: 675–678. A criticism of Formad's printed statements and conclusions concerning the etiology of tuberculosis. *New York Medical Journal* 40: 141–146, 172–177.

34. This statement is not completely correct. Gaffky and Loeffler were military officers but Koch was a civilian.

35. Now the Trudeau Institute, Saranac Lake, New York.

36. Trudeau, E.L. 1910. Some personal reminiscences of Robert Koch's two greatest achievements in tuberculosis. *Journal of the Outdoor Life* 7: 189–192.

37. Koch, 1883, loc. cit.

38. Privatbriefe von Robert Koch, *Deutsche Medizinische Wochenschrift* 37: 1443, 1911.

39. Koch, Robert, 1884. Die Aetiologie der Tuberkulose. *Mitteilungen aus dem Kaiserlichen Gesundheitsamte* 2: 1–88. Although completed in 1883, this paper was not published until 1884, when the second volume of the *Mitteilungen* of the Imperial Health Office appeared.

Chapter 15   *In Search of Cholera*

1. Koch, Robert. 1887. *Arbeiten aus der Kaiserlichen Gesundheitsamte*, Volume III, Julius Springer Verlag, Berlin, page 24.

2. The information in this section is based primarily on Wilson, G.S. and A.A. Miles, 1957, *Topley and Wilson's Principles of Bacteriology and Immunity, 4th Edition,* Volume II. Williams and Wilkins, Baltimore.

3. The information in this section is based primarily on *Arbeiten aus dem Kaiserlichen Gesundheitsamte,* Volume III, 1887. Julius Springer Verlag, Berlin. This volume, edited by Georg Gaffky with the collaboration of Robert Koch, provides exquisite detail on the German cholera expedition to Egypt and India. An excellent summary is also given in Möllers, op. cit. A brief summary in English, based primarily on Möllers, has been published by H. Mochmann and W. Köhler. 1983. *Indian Journal of Public Health* 27: 6–20.

4. Although nothing explicit was ever stated, it appears clear from a reading of the literature of the time that the fierce competition between Germany and France was partly responsible for the *rapidity* with which the German expedition was put together after the French announced their decision to go to Egypt.

5. If the term "German thoroughness" has any validity, it can certainly be applied to the *Mission Koch!*

6. Möllers, op. cit. page 138. See also *Deutsche Medizinische Wochenschrift* (1911), 37: 1443.

7. Anonymous. 1883. Dr. Koch's newly described cholera organisms. *British Medical Journal* 2: 828–829.

8. Möllers, op. cit., page 140.

9. A brief discussion of the French expedition is given by H. Mochmann and W. Köhler, 1984, *Meilensteine der Bakteriologie,* VEB Gustav Fischer Verlag, Jena, pages 171–173. See also Möllers, loc. cit., pages 621–622.

10. Straus, I., É. Roux, E.-I. É. Nocard, and L. Thuillier. 1884. Rapport sur le Choléra d'Egypte en 1883. *Arch. Physiol. Norm. Pathol.* series 3, page 381.

11. Frank, Robert M. and D. Wrotnowska (translators and editors). 1968. *Correspondence of Pasteur and Thuillier,* University of Alabama Press, University, Alabama, page 217.

12. Frank and Wrotnowska, loc. cit., page 223.

13. Letter from Émile Roux to Louis Pasteur dated 21 September 1883, sent from Alexandria. In Vallery-Radot, P. 1951. *Pasteur Correspondence,* Volume III, Flammarion, Paris, pages 398-399.

14. Howard-Jones, N. 1972. Choleranomalies: The unhistory of medicine as exemplified by cholera. *Perspectives in Biology and Medicine* 15: 422–433.

15. *Arbeiten der Kaiserlichen Gesundheitsamt,* 1887, loc. cit., page 153–154.

16. Möllers, op. cit., page 140.

17. Möllers, op. cit., pages 141–142.

18. Möllers, op. cit., pages 142–143.

19. Möllers, op. cit., page 144.

20. Mochmann and Köhler, loc. cit., pages 178–185.

21. Gaffky, *Arbeiten* loc. cit.

22. Arbeiten, loc. cit. Appendix, page 21.

23. Arbeiten, loc. cit. Appendix, page 23–24.

24. Anonymous. 1884. Dr. Koch's Sixth Cholera Report, *British Medical Journal,* March 22, 1884, pages 568–569.

25. Gläser, J.A. 1894. *Robert Koch's Komma-Bacillus ist nicht Ursache der Cholera. Urtheil eines ostindischen Arztes über die Ursachen (Aetiologie) der Cholera.* W. Mauke Söhne, Hamburg.

26. See footnote 3.

27. Möllers, op. cit. page 144.

28. Möllers, op. cit., pages 145–146.

29. Möllers, op. cit., page 147.

30. Hume, E.E. 1927. *Max von Pettenkofer.* Hoeber, New York.

31. In 1892, when Koch's cultures were available in Europe, Pettenkofer, 74 years of age, set out to *prove* that the cholera bacillus was not the cause of the disease by voluntarily swallowing a small quantity of a broth culture. In the presence of witnesses, he swallowed one milliliter of a broth culture containing about 10^9 vibrios. Cholera vibrios were isolated from his feces and he developed a mild, but apparently genuine, case of cholera (which he insisted, however, was *not* cholera). Some days later, Rudolf Emmerich, Pettenkofer's assistant, repeated this experiment and fell more severely ill, but recovered. See Breyer, H. 1980. *Max von Pettenkofer.* Hirzel Verlag, Leipzig. The culture was provided by Gaffky, who, having an idea why Pettenkofer wanted it, intentionally supplied a weakly virulent strain. See also Jordan, E.O. 1908. *A Textbook of General Bacteriology,* W.B. Saunders Co., Philadelphia, page 381.

32. The medal was a *Kronen-Orden II Klasse mit dem Stern.* Gaffky and Fischer were awarded *Rote Adler-Orden III Klasse.* Möllers, op. cit. page 153.

33. Möllers, op. cit. page 148.

34. Arbeiten, loc. cit.

35. Berliner Klinische Wochenschrift 1884, numbers 31, 32, and 32a.

36. British Medical Journal August 30, 1884, Volume II, pages 403–407 and 453–459.

37. British Medical Journal, Sept. 6, 1884, page 459.

Chapter 16   *The Pasteur/Koch Controversy*

1. Pasteur, Louis. 1882. De l'attenuation des virus. *Quatriéme Congrés International d'Hygiéne et de Démographie, Geneva* Volume I, pages 127–149. Koch's remarks are reported on page 145.

2. According to Bulloch, W. 1938. *The History of Bacteriology,* Oxford University Press, London, the term "microbe" was introduced by the French surgeon Charles Sédillot in 1878. Sédillot used the term "microbe" to refer to the whole body of infinitely small organisms, including vibrios, bacteria, bacteridia, etc. R. Vallery-Radot describes Sédillot's proposal movingly in his biography of Pasteur (*The Life of Pasteur*, 1937, Sun Dial Press, New York, pages 266–267). Pasteur heard Sédillot's address to the French Academy of Sciences in 1878 and adopted the term, making the whole world familiar with it.

3. Mollaret, H.H. 1983. Contribution à la connaissance des relations entre Koch et Pasteur. *NTM-Schriftenr. Gesch. Naturwis., Technik, Med., Leipzig* 20: 57–65.

4. Pasteur, Louis. 1877. Étude sur la maladie charbonneuse. *Comptes rendus de l'Académie des sciences* 84: 900–906.

5. Pasteur published many papers on anthrax during the years 1878, 1879, and 1880, consistently using Davaine's name for the bacterium, *bactéridie*, rather than Koch's term *Bacillus anthracis*. Indeed, in one footnote, Pasteur noted: "*Bacillus anthracis* of the Germans." In all, Pasteur published 31 papers on anthrax (mostly short), Koch only 2.

6. Koch, Robert. 1881. Zur Aetiologie des Milzbrandes. *Mitteilungen aus dem Kaiserlichen Gesundheitsamte* 1: 49–79.

7. Gaffky, Georg. 1881. Experimentell erzeugte Septicämie mit Rücksicht auf progressive Virulenz und accomodative Züchtung. *Mitteilungen aus dem Kaiserlichen Gesundheitsamte* 1: 80–133. Loeffler, Friedrich. 1881. Zur Immunitäts-frage. *Mitteilungen aus dem Kaiserlichen Gesundheitsamte* 1: 134–187.

8. Pasteur, Louis. 1881. Compte rendu sommaire des expériences faites a Pouilly-le-fort, près Melun, sur la vaccination charbonneuse. *Comptes rendus de l'Académie des sciences* 92: 1378–1383 (meeting of 13 June 1881). The experiment was done in May 1881 and the press and other visitors arrived to observe the results on 2 June 1881. The results of this experiment were widely reported in the newspapers of the day.

9. Frank, Robert M. and Denise Wrotnowska (editors and translators). 1968. *Correspondence of Pasteur and Thuillier concerning anthrax and swine fever vaccinations.* University of Alabama Press, University, Alabama.

10. Frank and Wrotnowska, loc. cit., page 111.

11. Judging from the reception which Koch's work received, and the prominence which Koch's laboratory was given at the German Hygiene Exposition (see Chapter 14), Thuillier's statement appears unlikely to be true.

12. Frank and Wrotnowska, loc. cit. pages 121–123.

13. Frank and Wrotnowska, loc. cit., page 127.

14. Pasteur, Louis. 1882. De l'attenuation des virus. *Quatrième Congrès International d'Hygiène et de Démographie, Genève* Volume I, pages 127–149. The attack on Koch is on page 128. Koch's brief rebuttal is reported on page 145.

15. Pasteur, loc. cit., page 145.

16. Vallery-Radot, Pasteur (editor). 1951. *Pasteur Correspondence.* Flammarion, Paris, page 313.

17. Vallery-Radot, loc. cit., page 314.

18. loc. cit, page 63.

19. Koch, Robert. 1882. *Über die Milzbrandimpfung. Eine Entgegnung auf den von Pasteur in Genf gehaltenen Vortrag.* Published as a separate pamphlet by Georg Thieme Verlag, Leipzig. See also *Gesammelte Werke* Volume I, pages 207–231.

20. Pasteur, Louis. 1882. La vaccination charbonneuse. Réponse à un mémoire de M. Koch. *Revue scientifique*, 3rd series V, 74–84.

21. Vallery-Radot, loc. cit., pages 430–431.

22. Vallery-Radot, loc. cit., page 431.

Chapter 17 *The Berlin Professor*

1. Prudden, T. Mitchell. 1886. On Koch's methods of studying the bacteria, particularly those causing Asiatic Cholera. *Eighth Annual Report of the State Board of Health of the State of Connecticut for the year ending November 1, 1885.* New Haven, 1886, pages 213–230. The writer was a pioneer American pathologist who was a professor at Columbia University. See *Biographical Sketches and Letters of T. Mitchell Prudden,* Yale University Press, New Haven, 1927.

2. One attendee at Koch's cholera course, the American William Henry Welch, later went on to become the dean of American medicine. Welch tells the story of how he prepared for himself a living culture of *Vibrio cholera* to take back with him to America. Subsequently in one of his lectures, Koch carefully explained why he had not brought a culture back from India, because Europe at that time was free of cholera. Hearing this story, Welch decided that since America was free of cholera, he should not bring his culture back. To avoid embarassment, he carefully sterilized the culture with sulfuric acid and poured it into the River Spree. Standing in the fog in the middle of the bridge, he saw his friend Carl Salomonsen from Denmark, another attendant at the course. Salomonsen, also moved by Koch's lecture, was disposing of *his* culture which he had intended to take back to Denmark.

3. Strangely, Robert Koch never had a bacterium named for him!

4. King, Lester S. 1952. Dr. Koch's postulates. *Journal of the History of Medicine* 7: 350–361. Lennox, John. 1985. Those deceptively simple postulates of Professor Robert Koch. *The American Biology Teacher* 47: 216–221. Doetsch, Raymond N. 1982. Henle and Koch's postulates. *ASM News* 48: 555–556. Carter, K.C. 1985. Koch's postulates in relation to the work of Jacob Henle and Edwin Klebs. *Medical History* 29: 353–374.

5. Koch's postulates, in modified form, apply not only to medical microbiology but to virtually any research study in microbial ecology. If one is interested in proving that a particular microbe is responsible for degrading a specific pesticide, for instance, one has to perform "Koch's postulates." Plant pathologists also discuss the "postulates" with great frequency.

6. Carter, loc. cit., page 354.

7. Koch, R. 1884. Die Aetiologie der Tuberculose. *Mittheilungen aus den Kaiserliche Gesundheitsamte* 2: 1–88.

8. It is often stated that Jacob Henle, one of Koch's teachers, first presented the postulates in a section of his book *Pathologische Untersuchungen* published in 1840. In a paper in this book entitled *Von den Miasmen und Kontagien* Henle formulates a germ theory of contagious diseases. "I will now adduce the reasons which prove that the matter of contagions is not only organic, but also animate, indeed, endowed with independent life, and that it can be thought of as a parasitic organism in the diseased body." He then goes on to show why this so-called *contagium animatum* is likely to be a living organism. Finally, he recognizes the importance of what would later be called cultural procedures: "In order to prove that they are really the causal material, it would be necessary to isolate the . . . contagious organism . . . and then observe especially the power of each one of these to see if they corresponded. This is an experiment which cannot be performed." Koch's contribution was to provide the methods so that such an experiment *could* be performed.

9. Loeffler, Friedrich. 1884. *Mittheilungen aus den Kaiserliche Gesundheitsamt,* Volume II. Although the volume as a whole is dated February 1884, Loeffler's article is dated December 1883. At this time, Koch was in India (see Chapter 15).

10. Koch, Robert. 1884. Die Aetiologie der Tuberkulose. *Mittheilungen aus den Kaiserlichen Gesundheitsamte* 2: 1–88. The quotation is from page 3. This article was written before Koch went abroad to study cholera.

11. See also Bulloch, William. 1938. *The History of Bacteriology.* Cambridge University Press, London. The reference to Koch's postulates is given on page 165.

12. Koch, Robert. 1884. Ueber die Cholerabakterien. *Deutsche Medizinische Wochenschrift,* Nr. 45. Also *Gesammelte Werken* 2: (1), 61–68.

13. Koch, loc. cit. page 68.

14. For a discussion of Pettenkofer's experiment on himself, see Chapter 15.

15. Wilson, G.S. and A.A. Miles. 1957. *Topley and Wilson's Principles of Bacteriology and Immunity,* 4th Edition. Williams and Wilkins, Baltimore, pages 1618–1619.

16. It is likely that in addition to "loyalty" to Prussia, Koch may have been reluctant to leave sophisticated Berlin for rustic Saxony.

17. See Möllers, op. cit., pages 167–187 for details on the Hygiene Laboratory.

18. Möllers, op. cit. page 174–175.

19. Martin Kirchner, as quoted in Möllers, op. cit. page 191.

20. Prudden, T. Mitchell. 1886. loc. cit. pages 213–230.

21. Russell, Harry. Diary, book 10, pages 63–64. University of Wisconsin Archives. Russell was in Koch's laboratory in 1890, just at the time of the tuberculin discovery (see Chapter 18).

22. The *Zeitschrift für Hygiene* was first published by Verlag Von Veit and Comp., Leipzig, in 1886. The first volume had 20 articles, dealing primarily with bacteriological studies having hygienic implications. This journal is still being published today but now has the English name *Medical Microbiology and Immunology.*

23. Möllers, op. cit. page 177–178.

24. Möllers, op. cit. page 179.

25. Möllers, op. cit. page 182.

26. Möllers, op. cit. page 185.

27. Möllers, op. cit. page 187. This book was written by Carl Flügge alone and went though a number of very successful editions. Flügge, C. 1886. *Die Mikroorganismen—Mit besonderer Berücksichtigung der Infectionskrankheiten.* Vogel, Leipzig.

28. Fraenkel, C. and R. Pfeiffer. 1889. *Mikrophotographischer Atlas der Bakterienkunde*, Berlin.

29. Among other things, Richard Pfeiffer discovered the phenomenon of immune bacteriolysis, an important basic discovery in immunology which also had practical implications for distinguishing pathogenic from nonpathogenic cholera vibrios.

30. Kirchner, Martin. 1925. *Robert Koch.* Julius Springer, Vienna.

Chapter 18 *Koch's Work on Tuberculin*

1. Doyle, A. Conan. 1890. Dr. Koch and his cure. *The Review of Reviews,* December 1890, 1: 552.

2. Koch, Robert. 1891. Ueber bacteriologische Forschung. *Verhandlungen des X Internationalen Medizinische Kongresses, Berlin, 1890.* Volume I, August Hirschwald, Berlin. Koch's address was given on August 4, 1890.

3. von Behring, Emil, and Shibasaburo Kitasato. 1890. Ueber das Zustandekommen der Diphtherie-Immunität und der Tetanus-Immunität bei Thieren. *Deutsche Medizinische Wochenschrift* 16: 1113–1114. This work, which will be discussed in the next chapter, was done in Koch's Hygiene Institute.

4. Details of the tuberculin reaction can be found in any medical microbiology textbook.

5. Anonymous. 1890. Editorial. *The Lancet* November 29, 1890, page 1168–1169.

6. The eminent Wisconsin bacteriologist Harry Russell was a visitor in Koch's laboratory in the fall of 1890, just after the tuberculin announcement. He states in his diary: "Koch announced his so-called cure for tuberculosis by means of tuberculin in Oct. [sic] 1890. Was *forced* to do so by pressure from the Emperor (Wilhelm II)." Volume 1 of Harry Russells diary, in the University of Wisconsin Archives, Madison, Wisconsin.

7. Koch, Robert. 1890. Weitere Mitteilungen über ein Heilmittel gegen Tuberkulose. *Deutsche Medizinische Wochenschrift*, November 15, 1890, 16: 1029–1032. The article was translated into English in a number of places, including the *British Medical Journal*, November 22, 1890, pages 1193–1195. Since Koch reacted to the tuberculin injection, it is likely that at one time he had been infected with tubercle bacilli.

8. Koch, Weitere Mitteilungen, 1890, loc. cit.

9. Koch, Weitere Mitteilungen, 1890, loc. cit.

10. Doyle, A. Conan. loc. cit., page 556.

11. Anonymous. 1890. General notes from Berlin. *British Medical Journal*, November 22, 1890, pages 1197–1198.

12. Doyle, A. Conan. loc. cit. page 549.

13. Gieson, G. 1981. Louis Pasteur. *Dictionary of Scientific Biography.* Scribners, New York, pages 350–416.

14. Anonymous. 1890. General notes from Berlin. *British Medical Journal*, November 22, 1890, pages 1197–1198.

15. Anonymous. Notes from Berlin. *British Medical Journal*, December 6, 1890, page 1327.

16. British Medical Journal, loc. cit. page 1241. This important principle of ownership of intellectual property has continued to this day in the academic institutions of most countries.

17. Official regulations for the sale of tuberculin were established in February 1891. *British Medical Journal*, February 28, 1891, page 485.

18. Doyle, A. Conan. loc. cit. page 556.

19. The Lancet, December 6, 1890, page 1241.

20. Möllers, op. cit., page 196.

21. Möllers, op. cit., page 196.

22. Russell, Harry, loc. cit.

23. Anonymous. 1890. Further researches by Professor Koch. Important statement by Sir Joseph Lister. *The Lancet*, December 6, 1890, 68: 1244.

24. It is very likely that Lister is referring to tetanus and diphtheria, the two diseases for which von Behring and Kitasato were about to announce the discovery of antitoxin. See Chapter 19.

25. Lister, Sir Joseph. 1890. Lecture on Koch's treatment of tuberculosis. *The Lancet*, December 13, 1890, 68: 1257–1259.

26. Trudeau, E.L. 1910. Some personal reminiscences of Robert Koch's two greatest achievements in tuberculosis. *Journal of the Outdoor Life* 7: 189–192. The quotation is on page 191.

27. Doyle does not seem to be aware that by this time Koch and his wife were virtually estranged.

28. Doyle, loc. cit., page 552.

29. The Lancet loc. cit., 1890, page 1291.

30. The Lancet, loc. cit., 1890, page 1294.

31. The Lancet, loc. cit., January, February, and March, 1891.

32. This summary, published in the *Klinisches Jahrbuch*, was translated into English and can be found in the *British Medical Journal*, March 14, 1891, pages 598–600.

33. *The Lancet*, loc. cit. March 14, 1891, page 631. There was one prominent tuberculosis patient whose cure was said to be due to tuberculin treatment: Paul Ehrlich. This is discussed in Ehrlich, P. 1913. Erinnerungen aus der Zeit der ätiologischen Tuberkuloseforschung. *Deutsch Medizinische Wochenschrift* 39: 2444–2446.

34. *British Medical Journal*, March 7, 1891, page 388.

35. Koch, Robert. 1891. Fortsetzung der Mitteilungen über ein Heilmittel gegen Tuberculose. *Deutsche Medizinische Wochenschrift* 17: 57–58. An English translation can be found in the *British Medical Journal*, January 17, 1891, pages 125–127.

36. Koch, Fortsetzung. loc. cit. page 57.

37. Koch, Robert. 1891. Weitere Mitteilung über das Tuberkulin. *Deutsche Medizinische Wochenschrift* 17: 1189.

38. *British Medical Journal*, February 14, 1891, page 374.

39. A vast number of charlatans capitalized on Koch's name, offering treatments at high prices. Some of the American examples are summarized in an article by S.A. Knopf, 1907, Quacks and quackeries. *Journal of the Outdoor Life* 3: 449–455. One quack was a physician with the name Dr. Edward Koch, who offered "Dr. Koch's Cure for Tuberculosis," advertising his cure alongside a picture of Robert Koch. Eventually, the Medical Society of New York exposed this fraud and drove him out of town. When Robert Koch was sent a copy of the newspaper article detailing this maneuver, he responded: "To my great satisfaction I read that at last some people have arisen to stop those quacks from continuing their nefarious work. . . . I would have long since brought suit against these swindlers, but friends familiar with the conditions in America advised against it, saying that, according to the law in the United States, these imposters could not be prosecuted. . . . Wishing you a complete success in your battle against these dangerous quacks, I remain, Yours, etc. ROBERT KOCH."

40. Möllers, loc. cit., page 201.

Chapter 19 *Consolidation and Transition*

1. Koch Robert. 1893. Wasserfiltration und Cholera. *Zeitschrift für Hygiene und Infektionskrankheiten* 14: 393–426.

2. Möllers, op. cit., page 206.

3. Böttger, P. 1891. Das Kochsche Institut für Infectionskrankheiten in Berlin. *Centralblatt der Bauverwaltung, 23 May 1891* 11: 201–203, 213–214, and 223–225. Provides considerable detail on the organization of the whole institute as well as on its construction, and provides plans of the buildings and the overall location. Another source of information on medical facilities in Berlin, including details regarding Koch's institute is *Berlin und seine Bauten*, 1896, Volumes I, II, and III. Wilhelm Ernst and Son, Berlin. The information on medical facilities is on pages 420–428 of Volume II.

4. Ludwig Brieger was the first person to suggest that pathogenic bacteria produce toxins and in 1887 presented evidence that the symptoms of tetanus were due to a toxin. His work thus formed the basis of Kitasato's important work on tetanus antitoxin. Brieger, L. 1887. Zur Kenntnis der Aetiologie des Wundstarrkrampfes nebst Bemerkungen über das Choleraroth. *Deutsche medizinische Wochenschrift* 13: 303–305.

5. Ackerknecht, E.H. 1953. *Rudolf Virchow. Doctor Statesman Anthropologist.* University of Wisconsin Press, Madison. This is apparently the only detailed biography of Virchow. It provides some interesting insights into this multifaceted man, one of Koch's antagonists and (sometime) supporter.

6. Excellent details of the history and organization of the Charité complex can be found in *Berlin und seine Bauten,* loc. cit.

7. Berlin und seine Bauten, loc. cit. It is amazing that in less than a year after the *first* announcement of tuberculin, Koch had not only obtained agreement for his new institute, but it had been built and occupied! Even in our "high-tech" age, such feats are rarely accomplished.

8. The details of the barracks arrangement were reported in *The Lancet* for August 29, 1891, page 517. There were nine barracks, seven for patients and two for attendants. The barracks for patients contained from 14 to 18 beds, and were so constructed that they could be kept scrupulously clean. "The inner surfaces of the walls ... are painted in oil with a coating of enamel, so that they can be washed at will with disinfectant. There are no little corners for dirt to collect in. The ventilation is so good that the air is completely changed twice very hour." We see here the beginnings of modern antiseptic practice.

9. Centralblatt der Bauverwaltung, loc.cit., page 224.

10. "So far as personal comfort is concerned, the furniture of these little laboratories is of Spartan simplicity, but their scientific equipment is most complete and select. The large incubators are built on the model of those in Pasteur's Institute, so that their internal temperature is not affected by the fluctuations of that of the surrounding air." *The Lancet,* August 29, 1891, page 518.

11. Möllers, op. cit., page 201–202. *British Medical Journal,* May 16, 1891, page 1096.

12. British Medical Journal, June 13, 1891, page 1295.

13. Möllers, op. cit., page 211.

14. The Institute of Hygiene eventually became the Institute of Medical Microbiology and moved its location to the building on Dorotheenstrasse that had housed Bois-Reymond's Physiological Institute, where Koch had given his famous first lecture on the tubercle bacillus (see Chapter 14). Although the building on Klosterstrasse was destroyed during World War II, the Dorotheenstrasse building, and the Institute of Medical Microbiology, are still extant. (The name Dorotheenstrasse was changed after World War II to Otto Grotewohl Strasse.)

15. Marquardt, Martha. 1951. *Paul Ehrlich.* Henry Schuman, New York. Bäumler, Ernst. 1984. *Paul Ehrlich. Scientist for Life.* Holmes and Meier, New York.

16. Behring's story has been told many times. For instance, H. Zeiss and R.

Bieling. 1940. *Behring. Gestalt und Werk.* Bruno Schultz Verlag, Berlin. A concise summary can be found in the book by Mochmann and Köhler, op. cit., pages 295–316.

17. von Behring, E. and F. Nissen. 1890. Ueber bacterienfeindliche Eigenschaften verschiedener Blutserumarten. Ein Beitrag zur Immunitätsfrage. *Zeitschrift für Hygiene und Infectionskrankheiten* 8: 412–433.

18. Behring, Emil, and Shibasaburo Kitasato. 1890. Ueber das Zustandekommen der Diphtherie-Immunität und der Tetanus-Immunität bei Thieren. *Deutsche Medizinische Wochenschrift* 16: 1113–1114. See Brock, T.D. *Milestones in Microbiology,* 1961, Prentice-Hall, Inc., Englewood Cliffs, NJ for an abridged English translation.

19. Behring and Kitasato, loc. cit.

20. Behring and Kitasato, loc. cit.

21. Behring, Emil. 1890. Untersuchungen über das Zustandekommen der Diphtherie-Immunität bei Thieren. *Deutsche Medizinische Wochenschrift* 16: 1145–1148. A portion of this paper is also translated in Brock, *Milestones in Microbiology,* loc. cit.

22. Behring, Emil and E. Wernicke. 1892. Ueber Immunisirung und Heilung von Versuchsthieren bei der Diphtherie. *Zeitschrift für Hygiene und Infectionskrankheiten.* 12: 10–44.

23. Marquardt, loc. cit., page 15.

24. Mochmann and Köhler, op. cit., pages 295–315.

25. Marquardt, loc. cit., page 35.

26. Koch Robert. 1893. loc. cit.

27. Koch, loc. cit. Although this work has a strong resemblance to that of John Snow in the 1850's in England, Koch never cites Snow and there is no indication that he ever read Snow's work.

28. Hazen, Allen. 1895. *The filtration of public water-supplies.* John Wiley, New York. An appendix on pages 139–143 of this book presents in English the rules of the German government regarding the filtration of surface waters for public water supply. In addition to Koch, members of the commission included Gaffky, Reincke, Flügge, and Fraenkel. The report of the commission was published in the *Zeitschrift für Gas-und Wasserversorgung,* 1894, page 185. See also the report of the cholera commission in Koch's collected works: *Gesammelte Werke,* Volume III, pages 873–875.

29. Jockisch, Hermann. 1939. Ein Brief Robert Kochs. *Deutsche Medizinische Wochenschrift* 65: 1609–1611.

30. This letter was published in Möllers, loc. cit., pages 208–209.

31. Many artifacts collected by Robert and Hedwig Koch during their extensive travels are preserved in the Robert-Koch Museum in the Institute of Medical Microbiology in East Berlin.

32. Bochalli, op. cit., page 112.

33. An excellent discussion of present-day tuberculosis in the light of Koch's work can be found in the Koch Centennial Memorial celebrating the 100th anniversary of Koch's discovery of the tubercle bacillus: Koch Centennial Memorial, *American Review of Respiratory Disease*, March, 1982, 125: 1–132.

Chapter 20   *Africa Years*

1. Salomonsen, Carl. 1914. Erringerungen an Breslau. *Berliner Klinische Wochenschrift*, March 16, 1914, page 486.

2. Möllers, op. cit., pages 265–266.

3. This was actually part of Koch's famous address (see Chapter 18) announcing the discovery of tuberculin, made at the International Medical Congress in August 1890. Über bakteriologische Forschung. *Verh. X. Internat. Congr. Berlin 1890*, Volume I, August Hirschwald Verlag, Berlin, pages 35–47.

4. Ade Ajayi, J.F. and Crowder, M. (editors). 1985. *Historical Atlas of Africa*. Cambridge University Press, Cambridge. This atlas provides an excellent overview of European colonization of Africa.

5. The year 1896 also brought the beginning of planning for a new Institute for Infectious Diseases, to replace the one that Koch had moved into in 1891 (see later in this chapter). The stay at the Charité location was terminated after such a short time because there was great need to expand the Charité Hospital and the only available land was in the vicinity of the Koch "barracks". However, planning and building the new Institute took at least four years, years in which Koch himself was primarily in Africa. This is discussed in detail in Möllers, op. cit., page 233.

6. Anonymous. 1896. Professor Koch's Visit to South Africa. *South African Medical Journal* 4: 211–214.

7. Anonymous, loc. cit., page 213.

8. Anonymous, loc. cit., page 212.

9. Koch, Robert. 1897. Researches into the cause of cattle plague. *British Medical Journal*, May 15, 1897, pages 1245–1246. Also, Koch, Robert. 1897. Prof. Robert Kochs Berichte über seine in Kimberley ausgeführten Experimentalstudien zur Bekämpfung der Rinderpest. *Deutsche Medizinische Wochenschrift* 23: 225–227 and 241–243.

10. Nicolle, M. and Adil-Bey. 1902. *Annales de l'Institut Pasteur* 16: 56.

11. Bochalli, op. cit., page 120.

12. Dwork, Deborah. 1983. Koch and the Colonial Office: 1902–1904. The Second South Africa Expedition. *NTM-Schriftenr. Gesch. Naturwiss., Technik, Med.*, Leipzig 20, 1: 67–74. Dwork's paper also has a short discussion of Koch's first South African expedition.

13. Koch, *British Medical Journal*, loc. cit., page 1245. Although it is not reported whether the South Africans were pleased with Koch's modest success, he was invited to Rhodesia in 1902, apparently because of the reputation he had gained during his earlier South African trip (see also Dwork, loc. cit.).

14. Meanwhile, the Russian bacteriologist Waldemar Haffkine, working at the Pasteur Institute, was considerably ahead of the German team and had already developed a method for the attenuation of the plague bacillus and prepared a vaccine. The German Commission tested and confirmed Haffkine's vaccine. Waksman, Selman A. 1964. *The Brilliant and Tragic Life of W.M.W. Haffkine, Bacteriologist.* Rutgers University Press, New Brunswick.

15. Koch reported in a letter from Bombay to his friend Libbertz that the temperatures in Bombay went as high as 50°C, and that for months the weather was 30°C and 70 percent humidity both day and night.

16. The Greek scientist Alexander Kartulis (1852–1920), an expert on amoebic dysentery, was a friend of Koch's from his cholera days in Egypt. The quotation in the letter is from Möllers, op. cit., page 235.

17. Research on malaria seems to have engendered violent controversies regarding priority, some of which Koch was involved in. The personal side of the malaria story, including Koch's involvement, has been well told in the following sources: Harrison, Gordon. 1978. *Mosquitoes, Malaria and Man: A History of the Hostilities Since 1880.* E.P. Dutton, New York; Ross, Ronald. 1923. *Memoirs, with a full account of the Great Malaria Problem and its Solution.* John Murray, London; Manson-Bahr, Philip. 1963. The Story of Malaria: The Drama and Actors. *International Review of Tropical Medicine* 2: 329–390. See also Foster, William D. *A History of Parasitology.* E. & S. Livingstone, Edinburgh. Most of the details of this controversy are ignored in the present book, as Koch was involved only peripherally.

18. Salomonsen, loc. cit.

19. They spent a few days in Naples awaiting their ship. For those who enjoy the delights of the Bay of Naples, the following passage from a letter Koch wrote to Pfeiffer may be interesting: "We're leaving at 7:00 o'clock this evening for the trip to the East Indies. We tried to spend a few days vacation at Capri before starting on this long, hot sea voyage, but the whole thing came a cropper. It was just as hot on Capri as at Naples, and we were so damnably plagued with mosquitoes that we couldn't stand it. And not a mosquito net in sight! After two absolutely sleepless nights we fled back to Naples, travelling across the bay right in the middle of a thunderstorm. Capri is ruined for me forever, and I'll never go back. However, scientifically the experience on Capri was very interesting and instructive, as we saw many places with mosquitoes with not a sign of malaria. Even the right species of mosquitoes. And now, we'll see what we can find in the East Indies." Möllers, op. cit., page 242.

20. German business interests had begun in the Pacific in Samoa, and then spread from there via Micronesia to Melanesia, through a company called *Deutsche Handels-und Plantagengesellschaft,* from which was formed in 1884 the *Neuguinea-Kompanie.* This company raised the German flag over Northeast New Guinea at about the same time that the British took over Southeast New Guinea. The western part of New Guinea was under Dutch control. The new German colony was called *Kaiser-Wilhelm-Land,* which from 1892–1899 was administered by the German Empire. After World War I the German colony went to Britain and was administered by Australia. The main harbor for the German colony was *Friedrich-Wilhelm Hafen*, now called Madang. The town of Stephansort, where Koch was based, was on one side of the bay. Geisler, Walter. 1930. *Australien und Ozeanien*. Bibliographisches Institut, Leipzig. The above information is on page 309.

21. Cairns, Australia established a memorial for Robert Koch in the center of the city, in honor of the man who freed the northeast Australian coast from malaria. Geisler, loc. cit.

22. Parts of New Guinea are geothermally quite active. While in New Guinea, Robert Koch had a geyser named for him! Möllers, op. cit., page 545.

23. Möllers, op. cit., page 253.

24. Koch, Robert. 1899. Ueber die Entwicklung der Malariaparasiten. *Zeitschrift für Hygiene und Infektionskrankheiten* 32: 1–24.

25. Koch, Robert. 1901. Address on malaria to the Congress at Eastbourne. *Journal of State Medicine*, No. 10. Koch's written and spoken English were by this time apparently quite good (see Chapter 21).

26. Koch, Eastbourne address, loc. cit.

27. Otto, R. 1960. Das Institut für Infektionskrankheiten "Robert Koch." *Forschungsinstitute, ihre Geschichte, Organisation und Ziele*, L. Brauer, A. Mendelsohn Bartholdy, and A. Meyer (editors). Paul Hartung Verlag, Hamburg, pages 3–10.

28. The clinical part of Koch's new institute was not actually completed until 1904, as part of the Virchow Hospital. In the meantime, the research was carried out at the new laboratory on Föhrerstrasse and the medical unit was established at the City Hospital at Moabit.

29. It is in this institute, now called the Robert Koch Institute, that Robert Koch's main archives are presently kept (see Figure 20.7). This building also contains the mausoleum which holds his remains. This building, in West Berlin, is one of two extant buildings associated with Robert Koch. The other building, in East Berlin, was the site of the Imperial Health Office (Luisenstrasse, now called Hermann-Matern Strasse; see Figure 9.2). The Institute of Infectious Diseases adjacent to the Charité (the Triangle Laboratory) and the building housing the Hygiene Institute were destroyed during World War II. There is also a Robert Koch Museum in East Berlin associated with the Institute of Medical Microbiology of the Humboldt University of Berlin, containing some artifacts donated by Hedwig Koch.

30. Gerber, Klaus. 1966. Bibliographie der Arbeiten aus dem ROBERT KOCH-INSTITUT 1891–1965. Special Publication of the *Zentralblatt für Bakteriologie, Parasitenkunde, Infektions Krankheiten und Hygiene, I. Abteilung Referate* 203: 1–274.

31. Koch, Robert. 1901. Die Bekämpfung der Tuberkulose unter Berücksichtigung der Erfahrungen, welche bei der erfolgreichen Bekämpfung anderer Infektionskrankheiten gemacht sind. *Deutsche Medizinische Wochenschrift* 27: 549–554.

32. Koch, Robert. 1903. Die Bekämpfung des Typhus. *Veröffentlichungen aus dem Gebiete des Militär-Sanitätswesens*, 1903, Volume XXI. August Hirschwald Verlag, Berlin.

33. Koch, loc. cit.

34. Dwork, loc. cit.

35. Dwork, loc. cit., page 70.

36. Möllers, op. cit., page 264.

37. Möllers, op. cit., page 264.

38. Möllers, op. cit., page 265.

39. Metchnikoff, Elie. 1939. *The Founders of Modern Medicine*, Walden Publications, New York, page 124.

40. The German versions can be found in Koch's Collected Works. The original English-language reports were sent to the colonial governments and also were published in 1903 and 1904 in the *Journal of Comparative Pathology and Therapeutics*, Volume XVI, pages 273–284, 390–397, and Volume XVII, pages 175–181. The research is briefly summarized in Dwork, loc. cit.

41. Möllers, op. cit., pages 265–266.

42. Möller, op. cit., pages 272–273.

43. Festschrift zum sechsigen Geburtstage von Robert Koch, herausgegeben von seinen Dankbaren Schülern, 1903, Gustav Fischer Verlag, Jena. The Preface to this volume bemoaned the fact that Koch was far away in Africa. The volume included research papers by 39 of Koch's former associates, including Ehrlich, Flügge, Fraenkel, Frosch, Gaffky, Kolle, Loeffler, Pfeiffer, Pfuhl, Proskauer, and Wassermann. The papers deal with almost all of the fields of bacteriology and immunology that had arisen out of Koch's fundamental work. His co-workers celebrated Koch's 60th birthday again on 23 July 1904 after his return from Africa, and still again on his 61st birthday on 11 December 1904.

44. Möllers, op. cit., pages 276–277. As we saw in Chapter 4, Koch's interest in archaeology began when he was still young.

45. Dispatch 30208 dated 8 August 1904 to the British Colonial Office, stored at the Public Records Office, Kew Gardens, London, as quoted by Dwork, loc. cit.

46. Dwork, loc. cit., page 73.

47. Foster, W.D. 1965. *A History of Parasitology*. E. & S. Livingstone Ltd., Edinburgh.

48. Möllers, op. cit., pages 306–307. Metchnikoff himself shared the Nobel Prize with Paul Ehrlich in 1908.

49. Möllers, op. cit., page 306.

Chapter 21   *Koch in America and Japan*

1. Anonymous. Robert Koch. *Journal of the Outdoor Life* 5: 164–169. The quotation giving Robert Koch's remarks is on page 169.

2. Loeffler, F. 1907. Zum 25jährigen Gedenktage der Entdeckung des Tuberkelbacillus. *Deutsche Medizinische Wochenschrift* 33: 449–451; 489–495.

3. Möllers, op. cit., page 328.

4. New York Times, April 8, 1908, page 8.

5. A copy of the menu and guest list are in the Archives of the American Society for Microbiology.

6. Anonymous. 1908. Robert Koch. *Journal of the Outdoor Life* 5: 164–169.

7. Anonymous, loc. cit., pages 165–166.

8. Anonymous, loc. cit., page 166.

9. Anonymous, loc. cit., page 167. The German orientation of the toast derived from the fact that the dinner was being held under the auspices of the German Medical Society of the City of New York. In a few years, World War I would change, for all time, American attitudes toward things German.

10. Anonymous, loc. cit., page 168.

11. Koch's speech was reported in English translation in loc. cit., page 169, and in German in the *New York Medical Journal*, 87: 748–749.

12. Winslow, C.-E. A. 1929. *The Life of Hermann M. Biggs.* Lea and Febiger, Philadelphia, pages 216–217.

13. Winslow, loc. cit., page 217.

14. Anonymous. 1932. Professor Robert Koch in Iowa in 1908. *Journal of the Iowa State Medical Society* 22: 153–154. There is also a letter in the archives of the Kitasato Institute in Tokyo which Koch wrote to Kitasato from Keystone. It adds little to this account.

15. Koch was mistaken. He expected to return to Europe from Japan by way of China, but he was ordered back to America.

16. Möllers, op. cit., page 347.

17. Now called the Kitasato Institute for Infectious Diseases. An excellent museum has many papers and artifacts from the association between Koch and Kitasato.

18. Möllers, op. cit., page 349.

19. Möllers, op. cit., page 350.

20. Letter to Geheimrat Martin Kirchner. Möllers, op. cit., page 350.

21. Maulitz, R.C. 1983. Robert Koch in the U.S.A. *NTM-Schriftenr. Gesch. Naturwiss., Technik, Med.*, Leipzig, 20: 80.

22. Theobald Smith was one of the most important early American pioneers in the field of bacteriology. His early work with the Bureau of Animal Industry of the U.S. Department of Agriculture was on Texas fever of cattle (a protozoal disease). Later, he isolated the causal agent of hog cholera, an organism which came to be named *Salmonella cholerae suis* (named in honor of Daniel Salmon, the chief of the department at the U.S.D.A. in which Smith worked).

23. Koch, Robert. 1908. The Relations of Human and Bovine Tuberculosis.

Journal of the American Medical Association 51: 1256–1258. Koch's paper is followed on pages 1258–1260 by comments by Theobald Smith and others.

24. Koch, loc. cit., page 1257.

25. Raw, Nathan. *Journal of the American Medical Association*, loc. cit., page 1259.

26. Koch, loc. cit., page 1258. This dogmatic statement appears to support those who contended that Koch had too high an opinion of himself.

27. Ravenel, M.P., Madison, Wisconsin. loc. cit., page 1259.

28. Conference in camera on human and bovine tuberculosis. *Journal of the American Medical Association* 51: 1262–1268.

29. Later in the conference, this prescription was rescinded, at the suggestion of Robert Koch himself.

30. Anonymous observer after Koch's theory regarding bovine tuberculosis had been roundly attacked at the International Tuberculosis Congress. Winslow, C.-E. A. 1929. loc. cit., page 218.

31. Smith, A.E. 1909. Report on the International Congress on Tuberculosis. *Illinois Medical Journal* 15: 303–310. The quotation is on page 309.

32. Smith, Theobald. 1932. Koch's views on the stability of species among bacteria. *Annals of Medical History*, N.S. 4: 524–530.

33. This is the Institute for Infectious Diseases in Föhrerstrasse (see Chapter 20). The mausoleum still exists and can be visited.

Chapter 22   *Assessment of Koch*

1. Ehrlich, Paul. 1910. Robert Koch. From an article in the *Franfurter Zeitung* which was reprinted in Himmelweit, F. (editor), *The Collected Papers of Paul Ehrlich*, Volume III, Pergamon Press, London, pages 601–610. The quotation is on page 607.

2. Ehrlich, loc. cit.

3. Dolman, Claude E. 1981. Koch, Heinrich Hermann Robert. *Dictionary of Scientific Biography*, Charles Scribners and Sons, New York, pages 420–435. The quotation is on page 430.

4. This has especially been the case in Germany, where the dominance of medical microbiology has been so strong that the general microbiologists formed for awhile their own organization affiliated with the American Society for Microbiology (the German Branch of the ASM).

5. Dubos and Dubos, op. cit.

6. Geison, Gerald L. 1981. Cohn, Ferdinand Julius. *Dictionary of Scientific Biography*, Charles Scribners and Sons, New York, pages 336–341.

7. Plesch, John. 1947. *János, The Story of a Doctor*. (Translated by Edward Fitzgerald). Victor Gollancz, London. The quotation is on page 50.

8. Nuttall, George H.F. 1924. Biographical notes bearing on Koch, Ehrlich,

Behring and Loeffler, with their portraits and letters from three of them. *Parasitology* 16: 214–223. The quotation is on pages 220–221.

9. Baldwin, Edward R. 1919. A call upon Robert Koch in his laboratory in 1902. *Journal of the Outdoor Life* 16: 1–2.

10. I am not able to confirm the story that Hedwig Freiburg was an actress. All German sources state that she was an artist, and that Koch met her when she was serving as the assistant to an artist who painted his portrait.

11. Metchnikoff, Elie. 1939. *The Founders of Modern Medicine. Pasteur. Koch. Lister.* New York, Walden Publications. The quotation is from pages 117–124.

Index

The author, Thomas D. Brock, standing in front of the Robert Koch Memorial in East Berlin, April 1987.